'This book offers a wide-ranging and timely critique of h.
is tackling the existential planetary crises facing humanity. It offers both complementary and contradictory insights on how key academics within a single university in the UK view progress, policy, and practice in implementing sustainability within higher education. What comes across clearly and urgently is the need for complementary and interdisciplinary based action across all sections of the higher education curriculum, and the book highlights how a critical learning transformation is needed in teaching practice and approaches to experiential learning in our universities. Otherwise, how can these centres of higher learning offer what can be categorised as a "good education?"'

Stephen Martin, *Honorary Professor in Learning for Sustainability at the Universities of Worcester and Nottingham, UK*

'We know that students at university today want to learn more about sustainability, and that they want it to be integrated into their subjects. This book offers inspiring examples of practice from different disciplines that demonstrate what good quality, integrated sustainability education can look like. It addresses key questions about the purpose of education and how transformative education can lead students to take action for social and environmental justice'.

Jamie Agombar, *Executive Director, SOS-UK*

'This is not your average textbook. One could describe it as a love letter to the paradox of sustainability education, calling out those who treat it as a bolt on to students' higher education experience. At its heart, this book is a collection of very personal perspectives of what "good" education means to and how it is expressed by dedicated academics at Canterbury Christ Church University who are striving to ensure that their students' experiences of higher education teaching and learning are purposeful. What is abundantly clear is the devotion and passion these individuals have, and they offer many wise words and share valuable concepts. A valuable read for both academics, as well as anyone who wants to better explore issues around the transformative potential of higher education as well as the "radical reimagining" of the University'.

Carolyn Hayles, *Professor of Environmental and Sustainable Design for the Built Environment, Cardiff Metropolitan University, UK*

'This book adeptly illustrates the complexity of the interplay between sustainability and education, contextualised within one university. It curates space to explore and reflect multiple, at times opposing, perspectives, transparently inviting the reader to relate, agree, disagree, and wonder. Impressively the collection manages to neither be prescriptive nor too general. It stimulates personal and professional reflection through its overarching paradox model with integrated examples of practice into theory. An inspiring read for practitioners at universities to deepen their understanding and practice of teaching in the 21st century'.

Kathrin Möbius, *EAUC – The Alliance for Sustainability Leadership in Education, UK*

'Bainbridge and Kemp, alongside a diverse collective of educators with multiple perspectives, offer a timely and pertinent book as well as a supportive guide to HE's crucial transitional task to "good" sustainability education embedded in sound practice. In courageously considering myriad concepts within this highly complex area, the book addresses issues often avoided for being too contentious and consequently too difficult. Tackling these issues is essential and through the authors' adoption of the Paradox Model, they succeed in provoking reaction and encouraging the reader to consider alternative approaches to adaptation. As a teacher educator in Higher Education, this book will provide a constant source of provocation and support now and well into the future'.

Tessa Willy, *Deputy Programme Lead – Primary PGCE, Institute for Education, UK*

Good Education in a Fragile World

This edited collection aims to provoke discussion around the most important question for contemporary higher education – what kind of education (in terms of purpose, pedagogy and policy) is needed to restore the health and wellbeing of the planet and ourselves now and for generations to come? The book contains contributions from colleagues at a single UK University, internationally recognised for its approach to sustainability education.

Introducing a conceptual framework called the 'Paradox Model', the book explores the tensions that underpin the challenge of developing sustainability in higher education in the 21st century. It asks probing questions about the purpose of higher education in the 21st century given growing concerns in relation to planetary safety and justice and calls for a rethinking of educational purpose. It draws upon the theory and practice of education and explores how these can develop an understanding of sustainability pedagogies in practice. Finally, it delivers thought-provoking discussion on what constitutes a 'good' higher education that meets the needs of a world in crisis. Drawing on a planetary health lens, the book concludes with a 'manifesto' that brings together the key insights from the contributing authors.

This will be an engaging volume for academics and educators from a wide range of disciplines in higher educational settings interested in translating sustainability theory into educational practice.

Alan Bainbridge is Reader in Education at Queen Margaret University, Edinburgh and Visiting Reader in Education and Sustainability at Canterbury Christ Church University.

Nicola Kemp is Reader in Education for Sustainable Futures at Canterbury Christ Church University.

Routledge Studies in Sustainable Development

This series uniquely brings together original and cutting-edge research on sustainable development. The books in this series tackle difficult and important issues in sustainable development including: values and ethics; sustainability in higher education; climate compatible development; resilience; capitalism and de-growth; sustainable urban development; gender and participation; and well-being.

Drawing on a wide range of disciplines, the series promotes interdisciplinary research for an international readership. The series was recommended in the *Guardian*'s suggested reads on development and the environment.

Interdisciplinary Perspectives on Socio-Ecological Challenges
Sustainable Transformations Globally and in the EU
Edited by Anders Siig Andersen, Henrik Hauggaard-Nielsen,
Thomas Budde Christensen and Lars Hulgaard

Sustainable Development Goal 16 and the Global Governance of Violence
Critical Reflections on the Uncertain Future of Peace
Edited by Timothy Donais, Alistair D. Edgar, and Kirsten Van Houten

Interdisciplinary Perspectives on Planetary Well-Being
Edited by Merja Elo, Jonne Hytönen, Sanna Karkulehto, Teea Kortetmäki,
Janne S. Kotiaho, Mikael Puurtinen, and Miikka Salo

Good Education in a Fragile World
The Value of a Collaborative and Contextualised Approach to Sustainability in Higher Education
Edited by Alan Bainbridge and Nicola Kemp

For more information about this series, please visit: www.routledge.com/Routledge-Studies-in-Sustainable-Development/book-series/RSSD

Good Education in a Fragile World

The Value of a Collaborative
and Contextualised Approach to
Sustainability in Higher Education

**Edited by Alan Bainbridge and
Nicola Kemp**

Routledge
Taylor & Francis Group

LONDON AND NEW YORK

earthscan
from Routledge

First published 2024
by Routledge
4 Park Square, Milton Park, Abingdon, Oxon OX14 4RN

and by Routledge
605 Third Avenue, New York, NY 10158

Routledge is an imprint of the Taylor & Francis Group, an informa business

British Library Cataloguing-in-Publication Data
A catalogue record for this book is available from the British Library

ISBN: 978-1-032-26097-6 (hbk)
ISBN: 978-1-032-26098-3 (pbk)
ISBN: 978-1-003-28651-6 (ebk)

DOI: 10.4324/9781003286516

Typeset in Times New Roman
by Newgen Publishing UK

Love is a precondition for any sustainable relationship (with anyone/ thing). That includes love for the dead, who, through the mystery of love, are in any case still alive.

Lise Cribbin
Roger Hayes
Yordanka Valkanova

We would like to welcome those who have recently joined us:

Ophelia Clough-Bainbridge
Trixie Ward

Contents

Figures

Contributors

Zulfi Ali is Senior Lecturer in the School of Humanities and Educational Studies at Canterbury Christ Church University. He has a background in international development, with over two decades of field experience in South Asia. His academic interests lie in the sociology and political economy of education policy and delivery in the postcolonial context, and in environmental and social justice education and activism.

Alan Bainbridge is Reader in Education at Queen Margaret University, Edinburgh, and Visiting Reader in Education and Sustainability at Canterbury Christ Church University. He has an expansive view of education that ranges across ecology, psychology, politics, and sustainability. He uses broadly narrative-based research methods to work qualitatively, exploring the development of education professional practice and the transformative nature of human/more-than-human relationships

Ronald Barnett is Emeritus Professor of Higher Education at University College London Faculty of Education and Society, and is the inaugural president of the Philosophy and Theory of Higher Education Society. He has written or edited more than 35 books on the philosophy and social theory of higher education, including *The Ecological University: A Feasible Utopia* and *The Philosophy of Higher Education: A Critical Introduction*. He is the inaugural recipient of the EAIR Award for 'Outstanding Contribution to Higher Education Research, Policy and Practice'.

Claire Bartram is Senior Lecturer specialising in Early Modern literature and culture at Canterbury Christ Church University. Her research focuses on the social history of writing, and publications include an edited collection of essays on *Kentish Book Culture: Writers, Archives, Libraries and Sociability 1400–1660* (2020). Her research interest in the socially enabling potential of education in early modern society intersects with this first foray into Critical University Studies, which draws on her role as recent convenor of the *Applied Humanities: Employability in Practice* module and current Section Lead for Humanities in the School of Humanities and Educational Studies.

Victoria Field is a writer and poetry therapist. Her most recent books are a poetry collection *A Speech of Birds* (2020) and a memoir, *Baggage: A Book of Leavings* (2016), She has published widely on therapeutic writing. She completed her doctoral studies on narratives of transformation in pilgrimage.

Diane Heath is a medieval cultural historian specializing in medieval gender, animality and animal studies and interested in regional and micro-history. She is Project Lead for NLHF Medieval Animals Heritage Project, and has a PhD in Medieval History from University of Kent, 2015.

Chiara Hewer is Careers and Employability Partner for the Faculty of Arts, Humanities and Education and the Student Enterprise Lead at Canterbury Christ Church University. She delivers support and consultancy to academic staff for the integration of Employability and Enterprise Education into the curriculum. She has led on the design and delivery of innovative employability modules, including *Applied Humanities: Employability in Practice*, across a number of undergraduate and postgraduate courses in the Faculty of Arts, Humanities and Education.

Nicola Kemp is Reader in Education for Sustainable Futures at Canterbury Christ Church. A geographer by background, her research interests are focused on the interrelationship between education and sustainability across ages and educational phases. Much of her research explores children's relationship with the natural environment. Her current research, funded by the Froebel Trust, focuses on developing outdoor provision for babies and toddlers and the potential of Nature Engaging and Nature Enhancing (NENE) pedagogies.

Ivan P. Khovacs did his PhD at the University of St Andrews on Theology and the Theatre, and teaches primarily in the areas of Christian Doctrine, Systematic Theology and Theology and the Arts. He also oversees Pastoral Theology and theological education in partnership with the Church of England. His research interests include theology and its intersections with theories of drama and the theatre, as well as film and the cinematic lens as theological analogues of seeing and being seen. Recent publications on practical theology and care of the environment bring together his work in Christian Ethics, Practical Theology, and Environmental Sustainability.

Sonia Overall is a writer, psychogeographer and Reader at Canterbury Christ Church University. Her publications include novels, poetry, short stories, academic articles and features, many of which explore place, the nonhuman, aspects of the weird and experimental forms. Her latest books are the pilgrimage memoir *Heavy Time* (2021) and the novel *Eden* (2022).

Alan Pagden has worked in education as a teacher and teacher educator since the early 1980s in the UK and abroad. He currently lectures at Canterbury Christ Church University specialising in primary education and education for sustainability. Alan's research interests include climate change education, global citizenship education and LINE (Learning in Natural Environments).

John-Paul Riordan is Senior Lecturer in Education at Canterbury Christ Church University in the UK. His research interests include video-based pedagogy analysis, science education, science 'misconceptions', conceptual change pedagogy, educational technology, and inclusion.

Stephen Scoffham is Visiting Reader in Sustainability and Education at Canterbury Christ Church University. A former teacher, president of the Geographical Association and school atlas consultant, he has written extensively about primary geography, global learning, creativity and environmental education. His latest books are *Sustainability Education: A Classroom* Guide (2022) and *Prioritizing Sustainability Education* (2020).

Peter Vujakovic is Emeritus Professor of Geography at Canterbury Christ Church University. He has been teaching students about sustainability since the 1980s. His teaching has been informed by his research on trees, from work on African savannas in his early career, to his recent interest in Green Heritage, with a specific focus on trees as 'gnarly agents' in burial grounds.

Tansy Watts is a lecturer on the early childhood studies degree at Canterbury Christ Church University and contributes to research exploring the contemporary relevance of holistic pedagogy.

Simon Wilson is Senior Lecturer in the Faculty of Arts, Humanities and Education at Canterbury Christ Church University, and a member of the Institute for Orthodox Christian Studies at Cambridge. He has a special interest in landscape, co-creation, love of learning, the theology of the Eastern Orthodox Church, and the true nature of sustainability. He is the author of 'The Fierce Urgency of Now and Forever and Until Ages of Ages: Study and the Restoration of Paradise on Earth', *International Journal of Sustainability in Higher Education* 23:1 (2022).

Foreword
To-ing and fro-ing – educating for a fragile world

The world is not merely in crisis *and* in conflicted motion, and in need of repair but – as so nicely captured in the title of this volume – is fragile. To say all this is at once to impose severe challenges on education, and especially on higher education. And the challenges come thick and fast, being both theoretical and practical. One person's sustainability is another person's assault on human rights, and provokes a severe counter. We see this in 'stop the oil' street protests which, even if peaceful, generate hostility from those immediately affected. In a nutshell, precisely because of the world's fragility, actions that are intended to be reparative may seem to make matters worse.

Globally, activists know this only too well. They see injurious practices – that are, say, harming the planet – and take action to mobilise public attention, knowing that they are stirring up a hornet's nest of counter-action from dominant forces defending their interests. Actual violence might ensue. Witness the hostility engendered by indigenous groups or ecological action groups that are trying to defend rain forests.

Fragility, therefore, goes all the way down. In an interconnected world, a fragility in one place helps to create fragility somewhere else. And so, simultaneously, we find fragility in Nature but also in geo-political relationships, in culture, in the economy, and in society more broadly. In turn, fragility is to be found deep in the heart of higher education, both in the claims made by students and professors (in a Chatbot age, can any utterance be trusted?) and in the teacher–student relationship, as teacher and student bring their different – and even conflicting – hopes into the learning situation.

A higher education for a fragile world, therefore, is confronted with the challenge of doing justice to a world that is fragile, and doing so with a pedagogical relationship that is itself fragile. And both considerations throw up yet further issues. The fragility of the world arises from its being a set of complex *and* interacting structures and forces: there is complexity and there is motion in the world. It is a world in which action and even – or especially – thought can have unforeseen consequences. It is, as some like to say, a world of emergence. Things happen; and all the more so in an Internet age. A thought, a text, spawns another, even from the other side of the world.

Such then, I take it, are some of the features of the background in forging a higher education for a fragile world. That preposition 'for' is potent since what is in question here, for this volume, is that of possibilities *for* the future that might be discerned in the emerging ecological age (as we may term it). This ecological age is presenting its challenges – of deformations in Nature and in all the other ecosystems of the world that bear upon Nature – but it is also presenting with possibilities afresh.

Such possibilities are not *there*, simply to be noted and for higher education then to set its goals accordingly with new sets of learning outcomes. No, these possibilities have to be discerned, imagined as feasible utopias, and taken on board in a redesign of the pedagogical setting. And this is where this volume scores so handsomely, in exploring possible pedagogies for this ecological age. In turn, these pedagogies would themselves become open-ended learning – and self-learning – situations, and highly collaborative at that. The discerning of possibilities would become part of the collaborative learning process.

These reflections presage the unfolding of a *fragile pedagogy* and in two senses. First, learning, especially in higher education, is not just fragile with students themselves being already in a fragile state. Understandably, students are in states of anxiety as they live their lives in a precarious world and have their eye on their futures, which are multiply precarious. Second, if a higher education is to be adequate to the precariousness of the world, it itself has to be *intentionally* precarious to some extent. Students have to be put into challenging situations to which they have to form their own responses. Open-endedness in the world requires open-endedness in higher education.

Another way of putting these reflections is to observe that a concern to bring about a world that is sustainable calls for a higher education that helps students to develop their own powers of *self-sustainability*. If the world is all the time presenting with new challenges, so pedagogies have to be both open-ended and challenging such that students can form their own powers to respond autonomously to unfolding situations. This requirement in turn calls for criticality in its fullest extent, where students – as graduates in the world – are able spontaneously to size up situations, form their own critical evaluations, and take actions that seek to restore harmony or at least mitigate perceived impairments in the settings in front of them.

There is here a to-ing and a fro-ing of sheer being in a difficult world, a world that is non-sustainable in its present form. Students – and then graduates – have to comprehend their situatedness in the world and their place in it. They have to go *to* the world, and seek to understand it, to reach out to the world, to listen to the world. This is the 'to-ing'. But they also have to withdraw from the world, in order to form their critical judgements of the world and to determine which critical actions that they may take and how. This is the 'fro-ing'.

'To-ing' and 'fro-ing': into and out of the world. This has to be the nature of a higher education for a world that is unsustainable, and not only in Nature but in all the human and all the other systems of the world. For example, to what extent are we equipping our students – most of whom will go into professions of some kind – with the powers and the human qualities to be whistle-blowers? Will they have the capacities to see matters in contexts, to form their critical evaluations of what they see around them to discern how their profession or organisation could–and should–be otherwise, and to determine a suitable course of action and have the requisite dispositions – of courage and so forth – to see it through?

It follows that a higher education that is concerned with sustainability in its fullest sense has to have some measure both of fragility but also of strength. Students' self-belief and courage and willingness to go on venturing forward and learning will be greatly aided by pedagogies having a high measure of collaboration, where students support each other and even affirm each other. This is not just a matter of the development of skills or even understanding but of the development of self, of having a centre to the self, some security in one's identity, fragile though that must be, too, in a fragile world.

All these matters and more can be glimpsed in the covers of this volume, so brilliantly orchestrated by Alan Bainbridge and Nicola Kemp. This is a text that is not just accessible in the way in which it has been curated but is carefully crafted as a conversation, inviting readers in, not least to continue the conversation in their own thoughts and practices. It has to be a never-ending conversation.

Ronald Barnett
London

Poems and provocations

Victoria Field

David Orr (2004, p. 94) writes that 'we experience nature mostly as sights, sounds, smells, touch, and tastes – as a medley of sensations that play upon us in complex ways'. Were education to reflect the world we inhabit, he says, we would have 'Departments of Sky, Landscape, Water, Wind, Sounds, Time, Animals and perhaps one of Ecstasy'. Poems evoke medleys of sensations and work, when they do, through the mystery and specificity of the material. They can illuminate the abstract and the general, but only through engagement with what Sonia Overall here describes as the hyperlocal, or what comparative theologian Mirabai Starr (2019) gathers under her capacious umbrella as the 'beloved'.

A poem, like a veil, both reveals and conceals the divine at the heart of everything. Everything matters. This iris, that sparrow, their child, your teacup, his shoes, the red trams of memory, the salt-slick taste of an oyster, our university. Academics argue about immanence and transcendence, but in a poem these work in tandem: in-breath, out-breath, left-foot, right-foot. A poem exists outside the chronic time explored by Lisa Baraitser (2017). It is recreated by the reader on every reading. It invites us to feel and respond. We may choose to look away, close our hearts and say nothing. We may choose to look.

References

Baraitser, L. (2017) *Enduring Time*. London: Bloomsbury Academic.
Orr, D.W. (2004) *Earth In Mind: On Education, Environment, and the Human Prospect*. London: Island Press.
Starr, M. (2019) *Wild Mercy: Living the Fierce and Tender Wisdom of the Women Mystics*. Louisville: Sounds True.

Acknowledgements

This book is deeply rooted
Rooted in time …
And in all the places, we have lived, worked and loved.
Seeds sown many years ago.

This is for you Lizzy – the one who first planted the paradox seed in my
consciousness and has been there to see it grow.

This is for Stephen – who's book, 'Using the School's Surroundings', over
forty years ago started my journey towards a 'good education'.

1 Introduction

Navigating educational tensions: The Paradox Model

Nicola Kemp and Alan Bainbridge

Our story of sustainability education

We want to start by outlining the origins and motivations behind the production of an edited text. Origins of ideas are always difficult to identify, but the story of how and why we have written it may be helpful in providing some context for readers – and the importance of context is central to our thinking about good education.

The book has been developed within one widening participation higher education institution located on a World Heritage site and with a Church of England foundation. Engagement with sustainability began in 2005, but it was not until 2011 when the University participated in the Higher Education Academy's 'Green Academy', and launched its curriculum development programme, *The Futures Initiative*, that – in the words of the Director of Sustainability, '*we had a beginning*'. Leadership in sustainability education was provided by a small Sustainability Team (comprising academic and support staff) with an aim of ensuring that 'every student has the opportunity to learn about sustainability in the context of their chosen discipline and field of work'. We received national and international recognition (Green Gown Awards 2017, 2018, Advance HE CATE Award 2020) for our approach toward sustainability education, which was based on four key principles that we will return to later in the chapter:

- Authentic collaborative partnerships
- Context sensitivity
- A critical and creative ethos
- Understanding the multidimensional nature of the University

The Paradox Model

Over this period, our interest developed in the relationship between education and sustainability at both a theoretical and practical levels. Situated within a Faculty of Education, we found ourselves applying educational thinking to sustainability, and vice versa, leading us to reject generic 'how to', or 'what

DOI: 10.4324/9781003286516-1

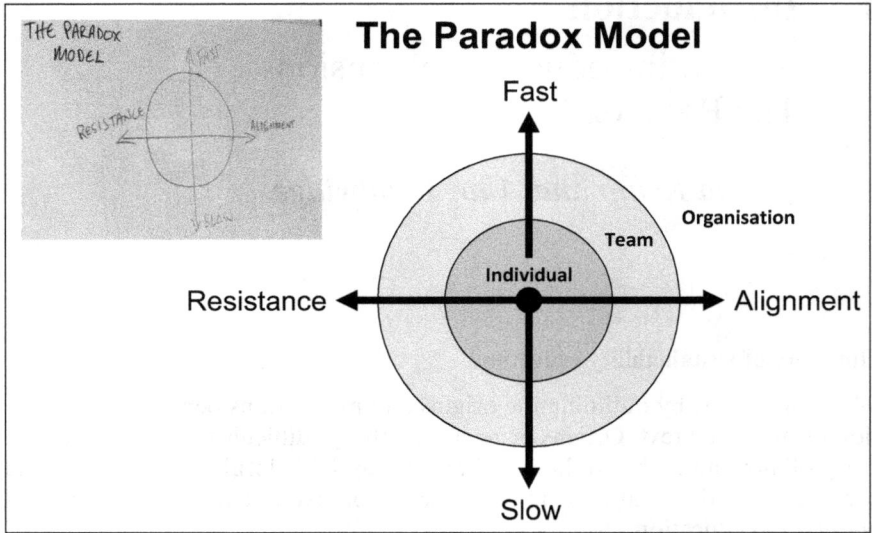

Figure 1.1 Paradox Model.

works' understandings of Education for Sustainable Development (ESD) with its suggestions of predefined pedagogical activities and content to be delivered. At the same time, the institutional commitment to embedding sustainability in the taught curriculum meant the demand from colleagues was for precisely such suggestions. As we sat in our office one day, we started to sketch our ideas on the whiteboard and identified two fundamental tensions within sustainability education that we felt were not being confronted within mainstream ESD research and practice (Figure 1.1). This was the start of our Paradox Model (Kemp & Scoffham, 2022) and we decided to make this the focus for the Sustainability in Higher Education (SHE) conference that we were due to host in Spring 2020.

The key principles of the Paradox Model were articulated in the conference call which is detailed below.

Those working within the context of higher education, are faced with the paradoxical challenge of addressing the urgent global crisis using the 'the slow way[s]' of education (Biesta, 2013, p.3).

A second challenge relates to the extent to which it is possible (or even desirable) to accept structural and systemic boundaries or whether as Biesta (2015) and others argue, educators have a 'duty to resist'. Universities themselves need to change; to push back against 'dominant ideologies' of neoliberalism which increasingly drive professional practice (Barnett, 2018:1). This raises questions about the extent to which drivers such as employability, quality, student experience and ranking offer useful opportunities to embed sustainability within university systems and structures or whether they need to be resisted.

These two challenges can be understood as paradoxes (fast/slow; alignment/resistance), which those working in sustainability in higher education must constantly navigate. We must act quickly in an environment that is slow to respond; we must challenge and change the very structures we are working within.

We propose a model in the form of a compass focused on these paradoxes. Its aim is to guide decision-making and identify when a particular type of response might be most appropriate. We argue that within the context of global crisis, professional wisdom is an increasingly essential virtue if universities are to be effective leaders in sustainability.

Whilst we were planning the conference and developing our thinking, the Covid-19 pandemic struck. The damage that has been 'done' to education is now well recognised. However, for us, the move to an online format opened up unexpected opportunities for colleagues and students from different institutions nationally and internationally to participate, increasing diversity and repositioning the conference as more outward facing. This apparent contradiction (that is, in locking down, opportunities can open up) was yet another example of the paradoxes within the contemporary higher educational landscape. Following a very successful conference, we continued to host monthly seminars to extend and develop our discussion and also guest-edited a special edition of the *International Journal of Sustainability in Higher Education* (Scoffham, Kemp and Consorte-McCrea, 2022).

Our focus: good education in a fragile world

For educators like us working in higher educational contexts, a fundamental question relates to what 'good' education means in the context of our fragile world (a theoretical concern) and how this impacts our professional practice (a practical concern). This book aims to provoke thoughtful discussion in what is a complex and contested arena, providing a point of departure to stimulate dialogue on what a 'good' education and a more hopeful future might be. It has been developed in ways that reflect the central principles of our approach to sustainability education.

An authentic collaborative partnership: The ideas contained within this book have been developed collaboratively with colleagues from across a diverse range of roles, disciplinary backgrounds, and life experiences. Our open call for chapter authors naturally created a diverse group (in terms of position, experience, and gender) who had not worked together before. We prioritised time and space to meet regularly over an extended period. This involved whole writing days and monthly lunchtime seminars led by chapter authors where evolving ideas could be discussed and debated.

Context sensitivity: The decision to focus on a single 'field of action' (in this case one university) reflects our belief that good education should be 'exquisitely sensitive to context' (Leithwood *et al.*, 1999, p. 4 cited in Bottery, 2016, p. viii). This context is purposefully reflected in the contributions, and we would encourage readers to reflect on the characteristics and opportunities afforded by their own context.

A critical and creative ethos: Although the contributing authors all clearly share a sense of urgency in this moment (i.e., the fragile), along with a commitment to turning education into a force for 'good', divergent (possibly irreconcilable) views are presented. We do not seek to present consensus – in fact it will be noticed that several of the contributions disagree with each other, in quite fundamental ways.

Reflective of the multidimensional nature of the University: We have purposely included contributions from a diverse range of colleagues including those with less experience of publishing and those from different disciplines. This is not accidental – it is a purposeful act that reflects the counter-cultural ethos of the book and recognises the need for all to engage with the question of what good education could look like in the contemporary higher education context. Equally, we feel it is important that sustainability education is not seen as something that can be simply 'added' to the existing curriculum offer. Instead, it requires working across disciplinary boundaries, negotiating meaning and developing shared understanding. It is difficult, uncertain, and disruptive. In doing so, we are seeking to illuminate the challenge from a diverse range of perspectives.

To provide perspective and focus for subsequent chapters, it is important to unpack two of the central concepts we are seeking to explore: fragility and goodness.

What do we mean by fragility?

The adjective fragile can be applied to an object that can be easily broken or destroyed, to indicate a tenuous relationship, weak expression, or be used as a constitutional descriptor. Our understanding reflects these diverse applications and in the following section, we briefly frame our understanding of fragility to situate our collective thinking.

Our fragile planet: The planet and its inhabitants are experiencing an existential crisis driven by human activities that have led to environmental crises, including climate breakdown; polar ice sheets melting at unprecedented rates; a global collapse in biodiversity, and the potential of a sixth mass extinction. When the list of crises is lengthened to include democracy, social care, technological advancement, and peace, it becomes increasingly difficult to hold in view the complexity of our predicament; humanity does indeed find itself living in a 'fragile world' of its own making.

The tenuousness of dominant understanding of the human/nature relationship: There are important ontological questions involving what it means to be human to be considered here; for example, what are the consequences of adopting certain philosophical stances vis-à-vis the nature-culture/human binary, and whether the 'ways of nature' are different to the 'ways of humans?' We can even question the possibility and potentiality of radical epistemological shifts, leading to the new ways of 'knowing' required to make the urgent 'turn' towards different values that are more aligned to human (and more-than-human) flourishing. These values would likely include intangible, untrainable qualities such as wisdom, integrity, intuition, reflexivity, and trust and, of course, we can question how such values could gain traction within contemporary higher educational culture.

The fragility, or weakness, of language: Fragility is also apparent in how we express and communicate the complex and often different understandings surrounding sustainability and education. Sustainability itself is a much-contested term and a rich and varied lexicon orbits the emerging discipline. We are sure that for some the struggle to put into words the complex relationships characteristic to sustainability debates will be seen as jargon intended to obfuscate. Perhaps time will determine which of 'wilding', 'rewilding', 'futures thinking', 'regenerative', 'de-growth', or even recent constructions such as 'Anthropocene', will be resilient. Educational language can be similarly problematic: Biesta (2013), for example, refers to 'learnification' to make the point that learning as a term has become over-used and is now applied to almost any context or situation without a clear sense of purpose – of the 'what' we are seeking to learn.

The fragility of our education systems: The corollary of Biesta's (2013) learnification can be witnessed in an education system unsure of its own purpose and, as several of the contributing authors argue, fragility may be inherent within Higher Education. Although it is fair to argue that most practitioners and academics working in education would oppose the current influence of neoliberal accountancy managerialism, the language of 'excellence', 'performance management', 'continuous improvement' and league tables remains dominant across the globe. That education is too fragile to resist managerial incursions, has led to what Hall (2021) refers to as 'The Hopeless University', where teachers and learners are continually disciplined and denied the possibility to follow their own educational desires.

Fragility can also be observed at the individual level and can perhaps represent the 'risk' of education. The educational encounter is not a dispassionate one, but more often than not a relational moment where a particular someone, learns a particular something, from another particular someone, for a particular reason. These are fragile moments where each participant is required to be open to new possibilities: the teacher imparts the knowledge about which they are passionate, casting it into an audience that, depending on their level of openness, may accept or reject such gifts.

What do we mean by 'good'?

There are multiple possible applications of the adjective 'good', so here we rehearse some of the thinking behind our use of the term. Exploring where 'goodness' might be experienced, we suggest, could help to find a way through the labyrinthine complexity of contemporary educational systems and to think about possible ways forward.

Good quality? The contemporary (we could argue neoliberal) language of higher education has become one of superlatives – of 'excellence', 'world-class', and 'continual improvement' where good is no longer good enough (Bainbridge, Gaitanidis and Chapman Hoult, 2018). Principles of competition are assumed to lead to improved professional practice, or educational attainment, in measurable ways that can then be used to publicly compare institutions and students. What is 'good' therefore has been claimed and anonymised by disembodied metrics and league tables, even to the extent that the idea of goodness makes moral claims, and the language of excellence can be interpreted as a distraction, or as avoidance of such claims (Bainbridge, Troppe and Bartley, 2022).

In contemporary educational discourse, good sits between the superlative 'outstanding', and the frankly dismissive 'requires improvement' in the OFSTED grading system. Whilst OFSTED only directly impacts universities that deliver teacher training, it shapes the schooling system that feeds into all HEIs and has established a culture of accountability, quality, and standards that permeate all mainstream educational settings. For universities, the Office for Student's Teaching Excellence Framework (TEF) uses similarly superlative language to promote a discourse of continual improvement. It 'aims to encourage higher education providers to improve and deliver excellence in the areas that students care about the most: teaching, learning and student outcomes' (OfS, 2023, p. 6). Questioning this dominant educational discourse of goodness allows us to think about new frameworks and processes for evaluating educational achievement and monitoring the accountability of professional practice. A responsibility of higher education can be to reinforce and celebrate learning that draws on unmeasurable priorities, principles, and values. We see higher education as a place of innovative and creative thinking, and where insights can be gained from integrating theory, research, and practice.

Good enough? The concept of good enough has its roots in psychology and, specifically in Winnicott's application, to mothering. In this book, we question whether the current neoliberal educational system is itself 'good enough'. Its limitations are increasingly recognised, not least in relation to addressing critical questions such as how humans can live on a finite planet without causing continuing catastrophic harm (Bainbridge, 2020). If education is to play at least a partial role in developing more sustainable futures, there are indications that the established aspects of dominant post-industrial educational systems are ill-suited to respond to planetary crises. Psychological insights can be applied to education that highlight the significance of relationships – between the learner and educator, between people and places. Good relationships can be the site where academic uncertainty, anxiety around not knowing and the assumed responsibility of the younger generation being handed over an increasingly depleted world, can be held and explored.

Inherently good? Theologically, the world as expressed in creation stories is understood as inherently good. For example, in the Christian tradition, the Bible repeatedly states, 'And God saw that it was good'. Christian scholars such as Caputo argue that creation can be understood as the movement 'from being to the good' (Biesta, 2013, p. 67). An orientation towards the 'good' in education can heighten our awareness of the significance of the places and spaces in which learning occurs. Opportunities for learning 'outdoors', at all ages and phases, are a recurring theme of this book. Questions can be asked in relation to the physical design of formal educational spaces, including the relationship between inside and outside worlds where principles and strategies of biophilic design could be used to create flexible, permeable spaces, to support learning, health, and well-being. It may also encourage us to take learning into places that may not be predictable but may be open to creative and individual exploration.

Outline of the book

We acknowledge that our focus on 'good education in a fragile world' requires staying with complex ideas but, to provide some supporting scaffolding for our readers, the book is organised into three main thematic sections:

(1) Rethinking educational purpose
   This section explores the micro to macro purposes of education, asking fundamental questions of education in the twenty-first century. Drawing on the Humanities, each chapter questions received narratives of what education and living sustainably might mean, offering pedagogic opportunities for students to locate themselves in the world and have agency to study and care for what they love.
(2) Pedagogies of (re)connection
   Pedagogies of (re)connection considers the complex relationship between the human and more-than-human other, as well as human/human interactions. The potential of both broad holistic approaches and detailed fine observation are presented, exploring how the individual is embodied within complex interconnected relationships, and how an appreciation of this provides learning opportunities towards a more sustainable future.
(3) Education as if the world mattered
   The final section provides three provocations towards imagining a higher education as if the world really mattered, each balancing between the imaginary and the acknowledgement of the reality of a sustainable future – as we type this, in May of 2023, the press informs us that the 1.5C climate threshold will be breached within four years (Harvey, 2023). Calls are made for a radical restructuring of societal conditions and educational process, for pedagogical approaches to embrace complexity and serendipity, while metaphorical Trojan Horse 'second operating systems' may be employed to counter the dysfunction of existing systems.

References

Bainbridge, A. (2020) Digging our own grave: A Marxian consideration of formal education as a destructive enterprise. *International Review of Education*, 66, 737–753 https://doi.org/10.1007/s11159-020-09866-7

Bainbridge, A., Gaitanidis, A. and Chapman Hoult, E (2018) When learning becomes a fetish: the pledge, turn and prestige of magic tricks. *Pedagogy, Culture and Society,* 26 (3), 345–361 DOI: 10.1080/14681366.2017.1403950

Bainbridge, A., Troppe, T., and Bartley, J. (2022) Responding to research evidence in Parliament: A case study on selective education policy. *Review of Education*, 10, e3335. https://doi.org/10.1002/rev3.3335

Barnett, R. (2018) *The Ecological University*, London: Routledge.

Biesta, G. (2013) *The Beautiful Risk of Education*, London: Routledge.

Biesta, G. (2015) The duty to resist: Redefining the basics for today's schools, *Research on Steiner Education,* 6 pp. 1–11.

Bottery, M. (2016) *Educational Leadership for a More Sustainable World*, London: Bloomsbury Academic.

Hall, R. (2021) *The Hopeless University: Intellectual Work at the End of the End of History.* Mayfly Books (available at: https://mayflybooks.org/wp-content/uploads/2021/05/The-Hopeless-University-E-book.pdf)

Harvey, F. (2023) 'World likely to breach 1.5C climate threshold by 2027, scientists warn'. *Guardian*, 17 May. Available at: www.theguardian.com/environment/2023/may/17/global-heating-climate-crisis-record-temperatures-wmo-research (Accessed 18 May 2023).

Kemp, N. and Scoffham, S. (2022) The paradox model: towards a conceptual framework for engaging with sustainability in higher education. *International Journal of Sustainability in Higher Education.* https://doi.org/10.1108/IJSHE-08-2020-0311

Leithwood, K., Jantzi, D., and Steinbach, R. (1999). *Changing Leadership for Changing Times*. Buckingham: Open University Press.

Office for Students (2023) *Teaching Excellence Framework Guidance*. Available at: www.officeforstudents.org.uk/media/7d4d14b1-8ba9-4154-b542-5390d81d703d/ra22-tef-framework-guidance-final_for_web.pdf

Scoffham, S., Kemp, N. and Consorte-McCrea, A. (2022) Guest editorial, *International Journal of Sustainability in Higher Education*, vol. 23 No. 1, pp. 1–3. https://doi.org/10.1108/IJSHE-01-2022-515

Roots

Victoria Field

You enter the main campus of Canterbury Christ Church University through a stone archway beside an ancient mulberry tree, possibly related to the one at the cathedral where, in 1170, the knights who murdered Thomas Becket left their swords.

The campus is built on the vineyards and gardens of St Augustine's Abbey which, in Saxon times, was the foremost centre of learning in Western Europe. The flinty remains of a wall that was part of the brewery and bakehouse still stand next to today's cycle racks. The university's courtyards are home to Kentish apples and hops, wassailing and beer-making, wild flowers and beehives, part and parcel of the garden heritage of Kent.

Up the road is St Martin's Church where King Ethelbert was baptised at the end of the sixth century. It's next to The Priory, also part of the university, which was never actually a priory but does have an authentic Tudor garden and a newish Cretan labyrinth mown into the grass near the rose beds. Things aren't quite what they seem.

The campus buildings date from 1962, and its distinctive chapel and tower block, along with the new medical school opened in 2022, are visible from miles around, but only from outside the city. Canterbury, with her dividing river, islands and thirty bridges, sits in a bowl surrounded by chalk hills. At street level, it's impossible to see how the city sits like a tilted egg in this corner of East Kent.

Acknowledgements: Parts of 'Roots' are taken from Field, V. (2020) *Not Utopia … But Maybe*, Canterbury: Canterbury Christ Church University.

DOI: 10.4324/9781003286516-2

Part 1

Rethinking educational purpose

The book starts by exploring the fundamental question of educational purpose in the twenty-first century. This section considers the micro to macro purposes of education, and each chapter offers pedagogic opportunities for students to locate themselves in the world and have agency to study what they love. The section offers a strong case for the value of the Humanities to challenge received narratives, to enable students to ask difficult questions, apply moral concerns, analyse language, and articulate who they are. The chapters reach for a richer understanding of the world that counteracts colonial, neo-liberalist, information-led, and managerialist approaches and values self-knowledge, an ethos of care and a vision of multiple possible futures in which our students are equipped to thrive.

DOI: 10.4324/9781003286516-3

2 'Swallowing a world'

Reflections on Education for Sustainable Development in higher education

Zulfi Ali

Swallowing a world

> Who what am I? My answer: I am the sum total of everything that went before me, of all I have been seen done, of everything done-to-me. I am everyone everything whose being-in-the-world affected was affected by mine. I am anything that happens after I've gone which would not have happened if I had not come ... I repeat for the last time: to understand me, you'll have to swallow a world.
>
> (Rushdie, 1981, p. 383)

The image of 'swallowing a world' as a way of understanding a person, a people, or even an idea is an intriguing one. In the quote above, Saleem Sanai, the protagonist of Salman Rushdie's novel *Midnight's Children* (Rushdie, 1981), challenges any illusions we may have about the ease of reaching high standards of knowledge and understanding, by suggesting that to understand him we would need to account for 'the sum-total of everything that went before' him, as well as that 'which would not have happened' had he not passed through the world. Just thinking about the sheer scale of the task involved in scrutinising all the causal chains that connect a person to the larger and ongoing historical trajectories involving pasts, presents, and futures is exhausting, if not impossible. It is a beguiling thought, and one that invites humility in terms of our claims of knowledge and understanding. The need to swallow a world, and the improbability of being able to do so, forces us to acknowledge the complexity of the worlds we create, while simultaneously positing severe limits on our epistemological ambitions. Recognising the high level of complexity involved in reaching a deep understanding requires us to acknowledge the shallow realms of very partial knowledge that form the basis of our actions. Or perhaps it invites us, in Rorty's tradition of the pragmatist theory of knowledge, to consider solidarity, rather than objectivity, as a starting point for making sense of the world (Rorty, 1989). In the absence of a 'God's eye view' (Putnam, 1981) or a 'mirror of nature' (Rorty, 2009) to reveal what reality really looks like, it points to the effort we need to apply and the complexities we encounter in understanding anything properly. It invites us to shun the

DOI: 10.4324/9781003286516-4

shallow, the predictable, the obvious, and the one-dimensional, in favour of complex, nuanced, and multidimensional insights into people, ideas, and the many realities we encounter. Whether we read Sanai's life as that of an individual, or whether we look at the narrative in *Midnight's Children* as a national allegory of postcolonial Indian history, the idea of the need to swallow a world to understand something is as powerful as it is insightful, even if we do not take it to its logical, literal extreme.

Theory, action, and theory-informed action

For well over two decades prior to joining academia, I worked in international development, mainly in South Asia, focusing on the many social educations – Education for Sustainable Development (ESD), human rights education, peace education, and development education. During this time, I found that while the work did have theoretical underpinnings, practitioners were often forced to think about ESD in very practical terms. When developing programmes and initiatives in a context that is at best challenging and at worst dangerous, the nuts and bolts of on-the-ground planning needed to be approached with deep pragmatism. The survival and flourishing of these programmes relied not just on strong praxis – theory-informed action – but also very practical contextual insights of what could work or not work in a specific site, in specific circumstances. So, on joining academia I welcomed the opportunity to take a step back and survey a larger theoretical canvas in trying to understand theory, action, and the interplay between theory and action in ESD, in any given context.

During my recent theoretical explorations, I started thinking about the idea of 'swallowing a world' as a pre-requisite to gaining a deeper understanding of the field and found the hyperbole of the idea useful as a framework for analysing the many social, cultural, political and economic challenges ESD faces in contemporary contexts. The idea that we need to swallow a world to understand the dynamics and tensions between theory and action in how ESD could be used to inform, inspire, and encourage into action generations of people, old and young, to help the planet pull back from impending catastrophe was an attractive one.

This chapter is an initial and rudimentary attempt at sharing some of my thoughts on this subject. It is a critique of the current state of ESD, but also an invitation to swallow a world to engage in a deeper and sharper analysis of the field that could lead to the start of many conversations about the future of ESD. My contention is that ESD today is too de-politicised, too de-radicalised and too disconnected from the social, political, economic, and other drivers that shape the world for it to be able to contribute significantly to the ongoing and urgent debates or actions involving environmental and social justice. On a planet staring into the abyss, educationalists seem to be pursuing 'business as usual' protocols, only occasionally and then also symbolically, threatening the status quo. This is particularly true even if we tentatively accept Read's view that our current civilisation is finished and something new must come out of it (Read and Samuel, 2019). My argument is that having acknowledged the severity of the existential crisis faced by humanity and reaffirmed a commitment to the belief that education must play a positive role – at best in averting this crisis and at worst in helping adapt to the new realities to be encountered – the field of ESD seems to be at a loss as to its contribution moving forward. An inability as well as an unwillingness as an educational community to 'swallow a world' places ESD in danger of becoming irrelevant.

In this chapter, I present preliminary historical sketches of four areas where I see gaps or failures in ESD, in the hope that these outlines will provoke a dialogue that will help explore the areas in more detail. One, a failure to critique the conventional narratives of international development which inform much of the understanding of sustainable development, and consequently approaches to ESD. Two, a failure to think beyond the narratives of free-market neoliberal capitalism when analysing the current political economy of the architecture of global decision-making, thereby severely restricting imaginations and visions of the future. Three, a failure to look at the world through lenses other than the Western, and to give due respect and consideration to other systems of thought, thereby narrowing the range of lenses or perspectives to view the world through to arrive at multiple, richer understandings. And four, using the slogan popularised by Greta Thunberg (Thunberg, 2019) and Extinction Rebellion (Extinction Rebellion, 2019), a failure to 'tell the truth' on environmental and social justice issues, partly due to the logic of promoting positive messages and avoiding mass environmental anxiety, but also due to the way higher education (HE) is organised and delivered.

Given the nature and scope of the chapter, I do not claim to present a comprehensive analysis of the field, nor do I pretend to be definitive in any sense. In fact, I leave the already substantial literature in this area to fill in the background of much of the technical detail about, for example, the multiple existential crises we face (WWF, CARE, and Actionaid, 2015; Grooten and Almond, 2018; IPCC, 2019; FAO, 2020) and the differences between Environmental Education (EE) and ESD (Hume and Barry, 2015). Instead, this chapter is based on a series of personal reflections from someone who has closely observed the developments in the field of ESD from the Global South.

The four failures

My central contention is that a large part of the problem lies in a failure to engage deeply with, and challenge, the prevailing mainstream orthodoxies, thereby resorting to a 'business as usual' approach that tries to find solutions in areas that are really the source of the problems. There seems to be an inability to see the problems holistically, or through a range of lenses. These structural constraints necessarily limit the ability to think critically and radically, thereby preventing the exploration of new paths. After all, one cannot pursue what one refuses to see. I will argue that this shrinkage of the circle of exploration manifests itself in at least four ways.

A failure to critique the conventional narratives on international development

The experience of living in a globalised world should highlight two relevant lessons. One is that environmental justice and social justice are deeply connected and cannot be seen in isolation from each other. The second is that environmental and social justice crises confront us simultaneously at the local as well as global levels. In a deeply connected planet, actions in one part of the world have consequences for everyone on the planet. The idea that environment and development are therefore intricately connected is not new. The very vocabulary of 'sustainable development' or 'Education for Sustainable Development' suggests a connection between ideas of environmental sustainability and social development. In fact, a very clear manifestation of the acknowledgment of this connection is that anchoring the ideas of ESD are the Sustainable Development Goals (SDGs). Like the 2000–2015 Millennium Development Goals (MDGs) before them, the SDGs (Sachs, 2012) in many ways have become our reference point for teaching and learning about ESD (NUS, 2019; EAUC, 2020; Longhurst and Kemp, 2021).

The problem, however, is that in the absence of an alternative framework for global discussion on these issues, the ESD literature takes a largely uncritical view of the SDGs. Despite the growing disillusionment with the SDGs and a recent open letter by academics asking the United Nations to drop them (Bendell, 2022), on the macro level, the SDGs remain the dominant discourse, uncontested and unchallenged in a significant way. Where critiques are raised, they are generally undertaken within the framework of the SDGs, aimed at certain themes, goals, or targets. For example, recently the SDGs have been critiqued for not having an impact on biodiversity conservation (Zeng et al., 2020), for a lack of clarity and precision in target setting (Vandemoortele, 2018), and for a commitment to endless growth (Eberth, Meyer and Heilen, 2023). The SDGs remain an integral part of global international development architecture, and that architecture is fundamentally flawed because of historical power imbalances. The world looks very different when viewed from the perspective of the marginalised in the Global South. These imbalances and asymmetries between the rich and poor countries are not new but have roots in colonial conquest. To buy into the SDG narrative is to buy into the narrative of the old colonial 'civilising mission' or the white man's burden, with a new lexicon based on international development aid. From Frantz Fanon in *The Wretched of the Earth* (Fanon, 1961), to Walter Rodney's *How Europe underdeveloped Africa* (Rodney, 2018) to Jason Hickel's *The Divide* (Hickel, 2017), there is a large body of literature that links colonialism, capitalism, and international development. The brutal extraction of resources under colonial rule not only made the coloniser rich and the colonised poor, but it also set the power relations between the two groups for the future. The SDGs, backed by narratives around development aid and helping the unfortunate catch up with the advanced West, are part of the old power relations, dressed up as benevolence and care. Yet, these critiques are largely ignored in ESD.

The SDGs, as is the case of all international development, are dominated by a group of Western controlled organisations – the World Bank, IMF, WTO – and the actions of these organisations have historically been far from benevolent (Stiglitz, 2003). Again, research shows that the global architecture of aid and development is designed to maintain the status quo of power relations (Hickel, 2017). By tracing the economic and policy histories of global development, Chang shows that the policy prescriptions, such as export subsidies, tariffs, and infant industry protections, used by all of today's developed economies are the complete opposite of the neo-liberal policies enforced on the developing world by the Washington Consensus (Chang, 2002, 2007). As Venn argues, the new forms of colonialism we witness today are more controlling and impactful than past modes (Venn, 2006). Of course, none of this was really unexpected and Albert Memmi had argued pessimistically back in the 1950s that it would be delusional to imagine that the colonial architecture of dominance would suddenly change after the wave of decolonising struggles (Memmi, 1957, 2006). Central to this debate is the western created 'myth of development', whereby

> they only perceive the economic epi-phenomena such as GDP growth, export performance, or the behaviour of the stock market; they do not notice the profound qualitative cultural, social, environmental and structural dysfunctions that prefigure the non-viability of the underdeveloped quasi nation-states in the new millennium.
>
> (De Rivero, 2001, p. 117)

Bypassing these complex debates based on the critiques of the SDGs specifically, and international development more broadly, has in my experience contributed to de-politicising and de-radicalising the field of ESD. It is not possible to talk about sustainable development without a historical understanding of the political economy of the global architecture of developmental decision-making, because power relations are central to the discussion. This means that to understand the field of ESD, we need to understand political economy, colonial and postcolonial power relations as well as sociology and anthropology. It really does entail 'swallowing a world'.

A failure to think beyond the narratives of free-market neoliberal capitalism

The global financial crisis of 2008 brought into sharp focus the political economy that dominates contemporary decision-making. Consequently, the literature attempting to make sense of those events, and the repercussions from them, took many forms: scholars and activists attempted to understand the build-up, the actual form of the financial crash as well as its fallouts (Varoufakis,

2015; Tooze, 2018); they celebrated the protest movements arising from it (Graeber, 2011, 2013; Chomsky, 2012); and they even called for a radical overhaul of how economics was taught (Chang and Aldred, 2014; Raworth, 2017). In the years immediately following the crisis, particularly around 2011 when the 'occupy' movement and the slogan 'we are the 99%', popularised by the anarchist anthropologist David Graeber collected momentum, there seemed real hope that the ideological orthodoxy at the root of the crisis, free-market neoliberalism, would be challenged and exposed. Yet, over a decade and a half after the financial crisis, neoliberalism continues to dominate the landscape of available ideas offered in terms of political economy. Four decades since its rise, neoliberal ideas remain powerful and dominant. The hold of this prevailing orthodoxy, what Bourdieu called the 'dominant discourse' (Bourdieu, 1998, p. 29) in political economy, of neoliberal capitalism as the undisputed solution to all our problems, is so embedded in our thinking that we cannot escape it. These ideas have been promoted with such coverage and intensity over time that they have come to be seen as natural, and indeed the only viable economic model for the flourishing of human civilisation. The failure to develop, articulate, and gain support for alternative and competing ideas, such as those articulated by the World Social Forum meetings under the slogan, 'Another World is Possible' (Fishes and Ponniah, 2003; George, 2004), has perhaps been the biggest failure of public imagination and activist efforts in recent history.

The destructive capacities of neoliberal capitalism, both in terms of environmental and social justice, have been convincingly documented and argued from a range of disciplines such as economics, sociology, politics, and anthropology (Chomsky, 1999; Bourdieu, 2003; Harvey, 2007; Roy, 2014; Hickel, 2017). These critiques have suggested that capitalism, which is entirely driven by the compulsion of growth and surplus creation, is incompatible with respect for nature and resources. Research also tells us that particularly in the past forty years, due to neoliberal policy prescriptions, inequalities have grown massively (Picketty, 2014; Alvaredo et al., 2018) and that such inequalities are incompatible with a nurturing and sustainable society and, indeed, world. As Klein reminds us in her book, *This Changes Everything*, capitalism is incompatible with sustainability (Klein, 2015). A good example for considering the intersections of neoliberal ideas permeating policies in international development is the case of the Structural Adjustment Programmes (SAPs), forced on developing nations by the International Monetary Fund (IMF). In the 1980s, over a decade, loans worth more than $28.5 billion were given to 64 countries in 187 projects under the umbrella of the SAPs (Reed, 2019). Even a cursory look at some of the critiques of SAPs, which include those from insiders at the World Bank and IMF, reveals a long and destructive history of interventions (Easterly, 2003, 2005). A large amount of documentation, analysis, and evidence suggests that instead of helping, the neoliberal free-market prescriptions of the SAPs have put the economies of large parts of Africa, South America, and South Asia on a downward trajectory (Hickel, 2017). Neoliberalism as an ideology has clearly made the rich richer and the poor poorer, just as its architecture has been deliberately designed to transfer wealth from the Global South to the Global North.

Yet, a sustained and deep critique of neoliberalism has largely bypassed the field of ESD. This necessarily makes any analysis severely restricted and compromised. Is it possible to understand the structural trajectories of the causes of environmental and social justice crises without challenging the ideologies that are largely responsible for these crises? Is it possible to teach and learn about sustainable futures while accepting a framework of ideas that are the very drivers of unsustainable practices? Unless the structural unsustainability built into the rules of global political economy is critiqued, exposed, and challenged, ESD will remain marginal to any move towards imagining and building sustainable futures. However, to do this the conversation needs to step outside the mainstream frame of reference of the SDGs. A world organised in different ways needs to be imagined. This would require freeing our imaginations and swallowing a world.

A failure to look at the world through epistemologies other than the Western-centric view

The invitation to swallow a world is an invitation to step outside a parochial and narrow perspective and to view the world in a more nuanced, empathic way. To begin building the frame of reference for ESD in the Global North with the Western, technological, scientific, industrial (or post-industrial), secular, and consumer society is understandable. However, the responsibility of educators is to ensure that space and time are provided to explore the world from different spiritual, social, and cultural lenses to build a deeper and more nuanced understanding. As Wade Davis, the Canadian anthropologist points out, compared to human history, the current Western perspective on the world is only three hundred years old, and that shallow history should not give us enough confidence to declare all other points of view invalid (Davis, 2009). He argues that the way we see the world, the narratives, and the stories we create to make sense of it, determine our relationship with nature. For example, he points out that in our current, profit-maximising worldview, trees exist to be chopped down (as long as they are a tree, they add nothing to the economy, whereas when they are cut and sold as timber, they add to the GDP), and mountains to extract ore from. We dig deep for oil, we frack for gas and, now that we are exhausting all possibilities, we are turning to tar sands, squeezing every last drop of oil out of particles of sand. Many indigenous communities on the other hand tell stories of trees and mountains having supernatural powers, as hosting spirits, as being their mother, their protector. They fear annoying nature. What is important is not who is right, says Davis. What matters is that one set of stories protects nature, and the other can only think of exploiting it to the maximum financial gain.

Ideas have consequences. We are now facing the consequences of a set of ideas based on exploitation, rampant consumerism, selfish individualism, and narrow, short-term self-interest. The hold of the scientific framing permeates and has consequences for every discipline, including the arts and humanities, putting, as Kundera suggests, 'the concrete world of life, *die Lebenswelt* ... beyond the horizon' (Kundera, 1986, p. 3). The positivist idea of science as an arbitrator of knowledge, coupled with the need to know everything means that we cannot relate to communities that do not have a desire for such knowledge. As Roy asks, '[W]hy can't we just be satisfied with not understanding ... To respect and revere the earth's secrets ... there ought to be a balance between curiosity, grace, humility and letting things be' (Barsamian and Roy, 2004, p. 19).

Following a long history of epistemicide (de Sousa Santos, 2015) and 'the disappeared knowledge systems' (Shiva, 1993), there is now a danger of heading towards a monoculture. The huge profits accrued from capitalist exploitation that drive international development also give us the confidence (or is it arrogance?) to consider Western worldviews as far superior to what we perceive as the superstitious, unscientific, primitive knowledge systems followed by traditional societies. As Shiva warns us, we are moving from wisdom to knowledge to (very partial) information, and in doing so we are losing touch with who we are and what our relationship is with the rest of the planet (Schooling the World, 2010). Ideas about indigenous knowledge being out of date, superstitious, not rational or scientific – ideas that are so deeply embedded in our thinking that we cannot seem to see beyond them. Start talking about this and you are accused of nostalgia, of romanticising old wisdom instead of moving ahead. As if old wisdom has nothing to contribute toward future perspectives. Such is our arrogance that, even though it is precisely these new perspectives that have pushed us to the environmental and planetary abyss, we still cannot see outside their frame.

There are thousands of communities with many thousands of voices that could enrich perspectives for ESD (Norberg-Hodge, 2000, 2003; Davis, 2009). There is an urgent need to talk more about these voices, listen to them and learn from them. More humility needs to be injected into ESD as a field. We may find a way to begin swallowing a world this way.

A failure to 'tell the truth' about social and environmental injustices

The fourth failure can be attributed to the modes of organisation and delivery structurally embedded in higher education (HE) that affect how ESD functions at different levels. To begin with a small example, the idea of 'swallowing a world' entails thinking in cross-disciplinary, multi-disciplinary, and trans-disciplinary ways. It could be argued that this demands a rigorous dialogue between environmentalists, earth and life scientists, engineers, sociologists, anthropologists, economists, geographers, historians, and educationalists to name only a few of the many academic divisions of knowledge we have created. Put in a stronger framing this could also mean that we need to come out of the silos of singular specialisms and individually become environmentalists, sociologists, anthropologists, economists, geographers, and historians to varying degrees. After all, these separations of knowledge into neat confines and divisions are human creations and, in the last four decades or so, organised largely on administrative and managerial lines, reflecting compulsions based on the ruthless logic of the supply/demand matrices in a market-driven HE sector. Despite attempts to create synergies between disciplines, contemporary pressures within HE make it difficult to do so. Additionally, the move that has taken place recently from the view of education as a social good to a commodity or an economic investment (Nixon, 2010; Collini, 2012, 2017), reduces the complexity of what goes on in higher education institutions and renders the goal of swallowing a world that much more out of reach.

At another level, the idea of swallowing a world and accounting for the past, present, and future in gaining understanding points to the causal chains that act as a 'cement of the universe' (Mackie, 1974). Historical links of action and consequence connect the past, present, and future, and bind us into sweeping historical trajectories as well as the contested interpretations of those trajectories. Sanai's use of 'a world' rather than 'the world' is significant here, for he understands that we live in multiple worlds, created and constructed by different peoples at different times. And who is to say whose world is the correct one? This opens the anthropological possibilities of a need for swallowing not just one, but multiple worlds. We are challenged to aim simultaneously for a long historical reading of the world and its inhabitants and to consider the many different readings and perspectives. All this points to the fact that history is important in ESD. Whether we are talking about planetary histories or civilisational histories or histories of ideas, historical readings can provide us with a deeper perspective. The problem is that, until very recently, such deep historical readings had gone out of fashion. David Armitage and Jo Gauldi point to the decline in the prestige and policy impact of history as a discipline as evidence to suggest that the way academic teaching and research are structured favours the consideration of shorter and shorter spans of history (Guldi and Armitage, 2014, 2015). They remind us of the tradition of the *'longue durée'*, proposed by Fernand Braudel and the French Annales School of History (Braudel, 1982, 2009), and advocate its return in analysing and understanding phenomena.

While there is some evidence of a revival of interest in larger spans of history, we still have a long way to go. The rise of postcolonial theory with the argument that you cannot understand postcolonial countries without appreciating the damage, both physical and psychological, inflicted during colonial occupation provides one glimpse of the changing winds (Memmi, 1957; Gandhi, 1998; Ahmad and Barsamian, 2000; Venn, 2006; Loomba, 2015). Yuval Noah Harari covers a sweep of history in response to the question about how it is that humans have moved from a position where homo sapiens were utterly insignificant in planetary terms to a position where there is a debate about whether we should rename our current epoch the Anthropocene (Crutzen et al., 2011), based on the scale of the impact human civilisation has had on the planet (Harari, 2014). Harari's answer, that this journey was made possible by the human ability to cooperate flexibly and in large numbers based on our powers of imagination, though stories and 'fictions' draw on deep history. The other recent example is Rutger Bregman's optimistic view of human beings and a hopeful history of their existence on this planet in his book *Humankind* (Bregman, 2020). Drawing references from a deep history of human beings, he offers perspectives of humans as 'friendly and peaceful' as opposed to the Hobbesian view of the brutish nature of life (Hobbes, 2017). In developing his argument, he too relies on large historical sweeps, drawing on anthropological, archaeological, political, and sociological ideas. The field of ESD still has a long way to go to incorporate long histories in its analysis.

Another aspect of the neoliberal influences on HE can be viewed in the way students are seen as consumers who need to be kept satisfied and happy. Like the mindfulness industry, one of the ancillary industries of neo-liberalism is the happiness or positivity industry. We are constantly reminded by influential thinkers like Bill Gates and Stephen Pinker to remain positive. The South Korean-born philosopher Byung-Chul Han calls our societies *positivity societies* and talks about pathological conditions deriving from an excess of positivity (Han, 2015). He argues that the twenty-first century is no longer Foucault's disciplinary society (Foucault, 1975), but an achievement society, turning us into subjects driven by the slogan 'yes we can'. There is no room for negativity or bad news in this society. As William Davis argues in the *Happiness Industry*, we have turned into narcissists (Davis, 2015). In such a society, bad news, harsh realities, and negative thoughts are to be avoided at all costs. This obviously narrows our field of vision and reduces the width and depth of our analysis. At a time when Greta Thunberg and the Extinction Rebellion are leading with calls to 'Tell the Truth' (Extinction Rebellion, 2019; Thunberg, 2019), this avoidance of the negative, based on the idea that reality is too scary and will put people off, has meant that ESD has often been reduced to feel good micro-level stories of fantastic, but small, initiatives. I am not suggesting that we should not celebrate local initiatives and great work. Only that this should not prevent us from telling the truth as an ESD community.

Some final thoughts

'Pessimism of the intellect, optimism of the will' was a phrase popularised by the Italian thinker Antonio Gramsci. Following Gramsci's advice, the ESD community needs to be brutally honest as intellectuals about the severity of the crises and the closing window for addressing these crises. But it must also promote hope by believing that if people worked together, change is possible, even in the worst possible scenario. Educators in ESD must either re-evaluate the role that they think education can play in the current crisis and therefore urgently undertake radically different and fresh ways of acting or admit that education does not have the power to create agency to face the crisis. Business as usual is not an option. To expect the same ideas, ideologies, and actions – whether they are economic, technological, or political – that have been tried for decades, to produce different results would be, in Einstein's definition, madness. This means that there is now a need to take an honest and difficult look at the factors that are hindering progress in ESD and find new ways of overcoming them.

If ESD has been deradicalized, depoliticised, and tamed in the neoliberal academy, efforts must be made to reverse these trends. Critical pedagogy can offer a way forward, for built into it is the idea of *praxis*, theory-informed action, so that education and activism are not separated. Critical pedagogy also helps us understand that a key purpose of education ought to be to ignite imaginations in radical ways so that different worlds and renewed possibilities can be imagined (Giroux and Franca, 2019).

Academia must take its responsibility for educating people, young and old, more seriously. As Pierre Bourdieu said:

> I have come to believe that those who have the good fortune to be able to devote their lives to the study of the social world cannot stand aside, neutral and indifferent, from the struggles in which the future of that world is at stake.
>
> (Bourdieu, 2003, p. 11)

References

Ahmad, E. and Barsamian, D. (2000) *Eqbal Ahmad, Confronting Empire: Interviews with David Barsamian*. Boston: South End Press.

Alvaredo, F., Chancel, L., Piketty, T., Saez, E., and Zucman, G. (2018). *World Inequality Report 2018*. Cambridge, MA, and London: The Belknap Press of Harvard University Press.

Barsamian, D. and Roy, A. (2004) *The Chequebook and the Cruise Missile: Conversations with Arundhati Roy: Interviews*. Boston: South End Press.

Bendell, J. (2022) *Initiative for Leadership and Sustainability: People will suffer more if professionals delude themselves about sustainable development – Letter to UN*. Available at: http://iflas.blogspot.com/2022/05/people-will-suffer-more-if.html (Accessed: 8 April 2023).

Bourdieu, P. (1998) *Acts of Resistance*. New York: New Press.

Bourdieu, P. (2003) *Firing Back: Against the Tyranny of the Market 2* (Vol. 2). London: Verso.

Braudel, F. (1982) *On History*. Chicago: University of Chicago Press

Braudel, F., and and Immanuel Wallerstein (2009) *History and the Social Sciences: The Longue Durée, Review (Fernand Braudel Center)*, 32(2), pp. 171–203. www.jstor.org/stable/40647704

Bregman, R. (2020) *Humankind: A Hopeful History*. London: Bloomsbury Publishing.

Chang, H.-J. (2002) *Kicking Away the Ladder: Development Strategy in Historical Perspective*. London: Anthem Press.

Chang, H.-J. (2007) *Bad Samaritans: Rich Nations, Poor Policies, and the Threat to the Developing World*. New York: Random House Business.

Chang, H.-J. and Aldred, J. (2014) 'After the crash, we need a revolution in the way we teach economics', *The Guardian*, 11 May. Available at: www.theguardian.com/busin ess/2014/may/11/after-crash-need-revolution-in-economics-teaching-chang-aldred (Accessed 8 April 2023).

Chomsky, N. (1999) *Profit Over People: Neoliberalism and Global Order*. New York: Seven Stories Press.

Chomsky, N. (2012) *Occupy*. London: Penguin.

Collini, S. (2012) *What Are Universities For?* London: Penguin.

Collini, S. (2017) *Speaking of Universities*. London: Verso Books.

Crutzen, P. et al. (2011) 'The Anthropocene: conceptual and historical perspectives', *Philosophical Transactions: Mathematical, Physical and Engineering Sciences*, 369(1938), pp. 842–867.

Davis, W. (2009) *The Wayfinders: Why Ancient Wisdom Matters in the Modern World*. Toronto: House of Anansi.

Davis, W. (2015) *The Happiness Industry*. New York: Verso.

Easterly, W. (2003) 'IMF and World Bank structural adjustment programs and poverty', *Managing Currency Crises in Emerging Markets*, pp. 361–392.

Easterly, W. (2005) 'What did structural adjustment adjust? The association of policies and growth with repeated IMF and World Bank adjustment loans', *Journal of Development Economics*, 76(1), pp. 1–22.

EAUC (2020) *The SDG Accord*. Available at: www.eauc.org.uk/the_sdg_accord (Accessed 8 April 2023).

Eberth, A., Meyer, C. and Heilen, L. (2023) ' "Economics for Future" from Different Perspectives – Critical Reflections on SDG 8 with a Special Focus on Economic Growth and Some Suggestions for Alternatives Pathways', in M.L. De Lázaro Torres and R. De Miguel González (eds.) *Sustainable Development Goals in Europe: A Geographical Approach*. Cham: Springer International, pp. 153–167. Available at: https://doi.org/ 10.1007/978-3-031-21614-5_8. (Accessed 8 April 2023).

Extinction Rebellion (2019) *This is Not a Drill: An Extinction Rebellion Handbook*. London: Penguin.

Fanon, F. (1961) *The Wretched of the Earth*. Reprint. London: Penguin Books, 1995.

FAO (2020) *State of Knowledge of Soil Biodiversity*. Available at: www.fao.org/3/cb192 8en/CB1928EN.pdf (Accessed 8 April 2023).

Fishes, W. and Ponniah, T. (2003) *Another World Is Possible*. London: Zed Books.

Foucault, M. (1975) *Discipline and Punish: The Birth of the Prison*. Reprint. Vintage, 2012.

Gandhi, L. (1998) *Postcolonial Theory: A Critical Introduction*. London: Allen & Unwin.

George, S. (2004) *Another World Is Possible If …* London: Verso.

Giroux, H. and Franca, J. (2019) All education is a struggle over what kind of future you want for young people, *CCCB Interviews*. Available at: https://lab.cccb.org/en/henry-giroux-those-arguing-that-education-should-be-neutral-are-really-arguing-for-a-version-of-education-in-which-nobody-is-accountable/ (Accessed 8 April 2023).

Graeber, D. (2011) 'Occupy Wall Street rediscovers the radical imagination'. *The Guardian*, 25 September. Available at: www.theguardian.com/commentisfree/cifamer ica/2011/sep/25/occupy-wall-street-protest (Accessed 8 April 2023).

Graeber, D. (2013) *The Democracy Project: A History, a Crisis, a Movement*. London: Penguin.

Grooten, M. and Almond, R. (2018) *Living Planet Report*. Available at: www.wwf.org. uk/sites/default/files/2018-10/LPR2018_Full%20Report.pdf (Accessed 8 April 2023).

Guldi, J. and Armitage, D. (2014) *The History Manifesto*. Cambridge: Cambridge University Press.

Guldi, J. and Armitage, D. (2015) 'The Return of the Longue Durée An Anglo-American Perspective', *Annales*, 70(2), pp. 219–247.

Han, B.-C. (2015) *The Burnout Society*. Stanford: Stanford University Press.

Harari, Y.N. (2014) *Sapiens: A Brief History of Humankind*. New York: Harper.

Harvey, D. (2007) *A Brief History of Neoliberalism*. Oxford: Oxford University Press.

Hickel, J. (2017) *The Divide: A Brief Guide to Global Inequality and its Solutions*. London: Penguin.

Hobbes, T. (2017) *Leviathan*. London: Penguin Classics.

Hume, T. and Barry, J. (2015) 'Environmental Education and Education for Sustainable Development', in *International Encyclopaedia of Social and Behavioural Sciences*, pp. 733–739.

IPCC (2019) *Global warming of 1.5 degrees*. Available at: www.ipcc.ch/sr15/ (Accessed 8 April 2023).

Klein, N. (2015) *This Changes Everything: Capitalism Vs. the Climate*. New York: Simon and Schuster.

Kundera, M. (1986) *The Art of the Novel*. London and Boston: Faber & Faber.

Longhurst, J. and Kemp, S. (2021) *Education for Sustainable Development Guidance*. Available at: https://media.www.kent.ac.uk/se/18355/education-for-sustainable-development-guidance2021.pdf (Accessed 8 April 2023).

Loomba, A. (2015) *Colonialism/Postcolonialism*. London: Routledge.

Mackie, J. (1974) *The Cement of the Universe – A Study of Causation*. Oxford: Oxford University Press.

Memmi, A. (1957) *The Colonizer and the Colonized*. Reprint. London: Routledge, 2013.

Memmi, A. (2006) *Decolonization and the Decolonized*. Minneapolis: University of Minnesota Press.

Nixon, J. (2010) *Higher Education and the Public Good: Imagining the University*. London: Bloomsbury Publishing.

Norberg-Hodge, H. (2000) *Ancient Futures: Learning from Ladakh*. London: Random House.

Norberg-Hodge, H. (2003) 'The consumer monoculture', *International Journal of Consumer Studies*, 27(4), pp. 258–260.

NUS (2019) *Student Opinion: Sustainable Development Goals*. Available at: https://uploads-ssl.webflow.com/6008334066c47be740656954/602e7b9c1d1ec10b6ca4f0a4_20190301_Student%20Opinion_SDGsreporting_FULL.pdf (Accessed 8 April 2023).

Piketty, T. (2014) *Capital in the Twenty-first Century*. Cambridge, MA: Harvard University Press.

Putnam, H. (1981) *Reason, Truth and History*. Cambridge: Cambridge University Press.

Raworth, K. (2017) *Doughnut Economics: Seven Ways to Think like a 21st-Century Economist*. Chelsea: VT: Chelsea Green Publishing.

Read, R. and Samuel, A. (2019) *This Civilisation is Finished: Conversations on the End of Empire-and What Lies Beyond*. Westerville: OH: Simplicity Institute.

Reed, D. (2019) *Structural Adjustment and the Environment*. London: Routledge.

De Rivero, O. (2001) *The Myth of Development: Non-viable Economies and the Crisis of Civilization*. London: Zed Books.

Rodney, W. (2018) *How Europe Underdeveloped Africa*. London and New York: Verso Trade.

Rorty, R. (1989) 'Solidarity or Objectivity', in M. Krausz (ed.) *Relativism: Interpretation and Confrontation.* Notre Dame: University of Notre Dame Press, pp. 167–183.

Rorty, R. (2009) *Philosophy and the Mirror of Nature.* Princeton: Princeton University Press.

Roy, A. (2014) *Capitalism: A Ghost Story.* Chicago: Haymarket Books.

Rushdie, S. (1981) *Midnight's Children.* Reprint. New York: Vintage, 1995.

Sachs, J. (2012) 'From Millennium Development Goals to Sustainable Development Goals', *Lancet*, 379, pp. 2206–2211.

Schooling the World (2010) *Directed by Carol Black [Film].* Lost People Films.

Shiva, V. (1993) *Monocultures of the Mind: Perspectives on Biodiversity and Biotechnology.* London: Palgrave Macmillan.

de Sousa Santos, B. (2015) *Epistemologies of the South: Justice Against Epistemicide.* London: Routledge.

Stiglitz, J. (2003) 'Democratizing the International Monetary Fund and the World Bank: Governance and Accountability', *Governance*, 16(1), pp. 111–139.

Thunberg, G. (2019) *No One Is Too Small to Make a Difference.* London: Penguin.

Tooze, A. (2018) *Crashed: How a Decade of Financial Crises Changed the World.* London: Penguin.

Vandemoortele, J. (2018) 'From simple-minded MDGs to muddle-headed SDGs', *Development Studies Research*, 5(1), pp. 83–89.

Varoufakis, Y. (2015) *The Global Minotaur: America, Europe and the Future of the Global Economy.* London: Zed Books.

Venn, C. (2006) *The Postcolonial Challenge: Towards Alternative Worlds.* London: Sage.

WWF, CARE and Actionaid (2015) *Loss and Damage: Climate Reality in the 21st Century.* Available at: wwfint.awsassets.panda.org/downloads/loss_and_damage___climate_reality_in_the_21st_century.pdf (Accessed 8 April 2023).

Zeng, Y. et al. (2020) 'Environmental destruction not avoided with the Sustainable Development Goals', *Nature Sustainability*, 3(10), pp. 795–798. https://doi.org/10.1038/s41893-020-0555-0

3 Education for life

Simon Wilson

Introduction

"What if education wasn't first and foremost about what we know, but about what we love?" asks James K.A. Smith (2009, p. 18). What if, indeed, education nurtured a knowing which is itself a form of loving, and a loving which is itself a form of knowing? What if knowing and loving were inextricably linked, and their intertwining enabled human persons and environment, or students and their subjects, to embrace each other, too, in an endless erotic[1] exchange? What if, at the very least, our education could cultivate in us the necessary predisposition for this dynamic loving relationship?

I have argued elsewhere (Wilson, 2020; Wilson, 2022) that an education of this sort would involve a return to the primary definition of the noun *study* as given in the *Oxford English Dictionary*: "Affection, friendliness, devotion to another's welfare; partisan sympathy; desire, inclination; pleasure or interest felt in something" (1989, p. 979). Study is "learning which leads to love and love which leads to learning" (Wilson, 2022, p. 29). However, it is not in itself a method, nor is it a system, a structure, or a design. It is immune to bullet-point teaching or handing out pre-digested droppings of data. A quality of attention to the world, to other people, to a subject studied, it is a practice of intense yearning or of desire that is in essence participative and poetic.[2] It is a love relationship. Through study, the eyes and ears are opened both to the beauty of the world and of whatever we happen to be learning (or teaching). We are then able to hear the call of beauty, which summons us out of ourselves back to ourselves, and into a world whose infinite depth is suddenly revealed (and being unfathomable, concealed at the same time). We gaze on our beloved and our beloved gazes back, and we and our beloved world will forever change as ever more depths are revealed.

So, our planet becomes once more a sort of Eden (a name meaning "delight": see Larchet, 2017, p. 36), and we save by surrendering any notion that somehow we *can* save or that we are even in control.

Sadly, love, beauty, delight, and the poetics of learning hardly seem fashionable subjects in education and may tend to be rejected as insubstantial, ephemeral, or even preciously effete by some sustainability advocates. As John

DOI: 10.4324/9781003286516-5

Zizioulas has written, "The ecological crisis we are facing seems to suggest fear – the fear of the destruction of our planet – as the prevailing motive for change of direction" (2021, p. 209). Such fear may have the energy to drive the speedy and heroic action some call for to save what we can of the planet; equally, it may lead to what Zizioulas calls "the managerial type of ecology" (2021, p. 226), which essentially views nature as something to be "to be managed, arranged, rearranged, distributed, and consumed" (2021, p. 145). The essential characteristic of both approaches is the idea that fear or dread of the future is to be overcome by an assertion of control.

The very concept of ecology may indeed be implicated in the managerial approach. As Ronald Barnett argues,

> The idea of ecology points to the interconnectedness of all things in the world (and even beyond). It gains its power from a sense of impairment, especially a sense that systems – ecosystems – have been corrupted in some way.
>
> (2018, p. 17)

Ecology thus views the natural world as a system that, no matter how complex or inchoate it may be, can be subjected to management. Its abstractness is revealed in the language Barnett uses, which conceives of interconnectedness rather than deep personal relationships.

Love talk seems to have little place in any of these serious ventures: love needs the particular, the unique, the specific if it is to exist, as indeed does the world.

Behind this turn to *study* is an ancient (in this case fourth-century) yet crucial insight: "Human nature came into being as something capable of becoming whatever it determines upon, and to whatever goal the thrust of its choice leads it" (Gregory of Nyssa, 2012, p. 113). In other words, what we desire, and what we desire to know, shapes us (cf. Carnes, 2014, pp. 200–201). If we determine upon a fear-driven approach, whether with fantasies of last-minute heroism or systems control by teched-up managers, we may end up in a world in which love is rendered marginal, unimportant, perhaps even irrelevant to the struggle ahead, a waning thing in a world of abstracts and supercomplex systems. That would be disastrous as, in the words of Kallistos Ware, "love is the only true answer to our ecological crisis, for we cannot save what we do not love" (2013, p. 105).

By contrast, it is the argument of this chapter that unless we *study* in education, unless a mutual erotic yearning for beauty is cultivated in education, human beings as fully human beings and the world as world in the fullest sense have no future. Our specific concern here, then, is with the relationship to the future – of the planet and of individual persons – as envisaged by *study*: an infinite stretching out of ever more blissful loving knowledge, as we follow the call of beauty, not that of fear or apprehension or control or indeed safety. We cannot know where it will lead us.

The sublime and the beautiful

Perhaps we need to understand why beauty, like love, has become a rather unfashionable subject in academia. Once, for the ancient Greeks, it was "a phenomenon worthy of the most rigorous intellectual inquiry" (Sammon, 2017, p. 18). It inspired virtue and uplifted one towards the divine, and it did that by evoking love (Carnes, 2014, pp. 17–18; Sammon, 2017, pp. 18–21). But now beauty has been marginalised, reduced to a product or something that can be managed with the correct technology. It has become a commodity; something to be bought and sold. Beauty is the most frivolous department in a store, "the realm of cosmetics, the ornamental, the decorative" (Carnes, 2014, p. 25). Serious people (perhaps specifically serious *men*) roll their eyes and shuffle their feet with discomfort in the Beauty section.

This marginalisation of beauty can be understood as being the result of two main developments. The first is the rise of the aesthetic in the eighteenth century (see Carnes, 2014, pp. 17–27). The aesthetic effectively reduces the elevating and transformative meaning of beauty to fleshly pleasures and trivial or even unhealthy pursuits. Here, the single most-influential account in the English language was undoubtedly Edmund Burke's *A Philosophical Enquiry into the Origin of our Ideas of the Sublime and Beautiful* (1759). While Burke associates beauty with love, his treatise also insists its pleasantly relaxing qualities can lead to languor, lassitude, and enfeeblement (1998, pp. 177–178; Carnes, 2014, p. 20). This attractive yet decadent world of beauty, with its attendant "relaxing [of] the solids of the whole system" (Burke, 1998, p. 177), is explicitly gendered feminine (Eagleton, 1990, pp. 54–55).

At the same time as linking beauty to effeminate and dangerously soft flesh, eighteenth-century aesthetics further sideline it by promoting the sublime in its place (Carnes, 2014, pp. 1727). Burke describes the sublime as a kind of modified terror (1998, pp. 102), an antidote to love, which is necessary to "exercise the finer parts of the system" (ibid. p. 165). Without the masculine discipline of the sublime, we would evidently fall into effeminacy and the embarrassment of relaxed solids, leading to dissipation both of body and mind.

As beauty subsided, it was replaced by the sublime as the mark of the highest of human experience and knowledge (Carnes, 2014, pp. 30–32). When Rudolf Otto came to identify the essence of religion in an experience of the numinous, with all its awe and terror, he remarked on the "hidden kinship between the numinous and the sublime" (1958, p. 63). Consequently, "the holy looks remarkably like the sublime" (Carnes, 2014, p. 31). Later, despite their lofty scepticism and even nihilism, many of the central thinkers of what may be loosely termed postmodernism found the sublime a much more congenial topic than the beautiful (Hart, 2003, pp. 43–93). Beauty and love, it seems, are not worth talking about (unless subjected to the scrupulous discipline of sublime terror, in which case they effectively cease to be).

We may be said to live in an age in which the sublime continues to rule. The future of the planet, fantasised as being a prolongation of terror and pain, is there to rein us in. Education institutions themselves are deemed flagrantly flabby and subjected to the sublime regimen of managerialism:

> Managerialism claims to restructure bureaucratic organizations for greater efficiency and economy. These claims are linked to the belief that the public sector is inefficient and wasteful, and thus not giving value for money because of the absence of an automatic disciplining mechanism.
>
> (Lea, 2011, p. 817)

Studying and teaching are controlled by the firm hand of outcomes, quality control, and constant evaluation (Lea, 2011), all of which attempt to control and coerce the academic, the student, and the very subject studied. Education has thus become an exercise in the managerial sublime, whose rigours reduce the call of beauty, if it is heard at all, to a tight strangulated squeak. Study, in the sense outlined above, has little place here, and humans adapt themselves to what is desired, transforming themselves into Lea's "automatic disciplining mechanism[s]" (2011, p. 817). So, in the words of Richard Hall, the university "functions as an anxiety machine" which denies "that for which humans yearn" (2021, pp. 2–3). As a concomitant, it manufactures hopelessness about the future (Hall, 2021). Thus, the managerial sublime, as imposed on academics and students, serves to sever us from any sustainability, whether ecological or educational.

Beauty restored

There is, however, an alternative, if relatively little known, discourse on beauty that regards it once more as a revelation of infinite potential and transform-ation, something which could liberate the yearning of the studious. This is to be found in theology. Notable recent examples include Hart (2003), Sammon (2013, 2017), Carnes (2014), and Johnson (2020), all of whom owe a greater or lesser debt to von Balthasar's great seven-volume *The Glory of the Lord* (1982– 1989). The present chapter rests to a certain extent on the insights of these authors but, in particular, on insights offered by the theology of the Eastern Orthodox Church. It is not, however, an *apologia* for a specifically Christian sustainability education, and does not require a particular religious outlook. Rather it concentrates on beauty, love, and poetry as experiences that require no investment in religious belief. Study, then, does not in any way require a church-based education, but is rather beauty- or perhaps poetry-based.

In this context, it is worth remembering the main reason given in Lynn White Jr's still influential "The Historical Roots of Our Ecologic Crisis" (1967) for explicitly excluding the Eastern Church from accusations that Christianity is radically implicated in environmental catastrophe. White writes that nature is conceived by Orthodoxy "primarily as a symbolic system through which God speaks to men ... This view of nature [is] essentially artistic rather than scientific" (p. 1206). As Chryssavgis and Foltz comment,

White's understanding of Orthodox Christianity here is far from flawless. But his instincts are sound, for he sees that in the Christian East … creation was understood not as an impersonally functioning mechanism but aesthetically … as an endless love poem … meant for us to read and reread, to heed and treasure.

(2013, p. 3)

Elsewhere, Foltz argues that it was early Christianity that first cultivated the very love for nature, the "feel for nature," the appreciation of the beauty of nature for its own sake, beauty that not only inspired the romantic movement of the nineteenth century, but continues to animate the environmentalism today (2014, p. 235).

Basil the Great, in the fourth century, for example, wrote that "the world is a work of art displayed for the beholding of all people" (St Basil the Great, 1978, p. 55). It was the same Basil who is credited with composing "what is likely the first example in antiquity of praise for a natural beauty that is not pastoral, but wild and uncultivated" (Foltz, 2014, p. 237). Given even half a chance, the beauty of this work of art, this infinite love poem, may still attract the desire of the studious heart, despite the disciplinary efforts of the sublime in their various manifestations, and thus liberate us from our internalised disciplining mechanisms.

The philosopher, poet, literary critic, and theologian Vladimir Soloviev (1853–1900) describes how this liberation may occur. Humans, studying the beauty of creatures, participate in that beauty, are filled with it, until they complete it in artistic creation, where it has a perfection not found even in nature:

[T]oday's art, in its greatest works, captures flashes of eternal beauty in our current reality and extends them further, forewarns, gives presentiment of a supernatural future reality for us, and serves, in this way, as a transition and connecting link between the beauty of nature and the beauty of the world to come. Art understood in this way ceases being empty amusement and becomes an important and edifying concern,… in the sense of inspired prophecy.

(2020, p. 75)

Soloviev's artist is in effect a student. She concentrates her pleasure, her desire, and her sympathy on the beauty of animals, which enables her to carry further nature's hints and yearnings for the perfection of beauty. The artist/student captures a timeless yet ever-shifting moment that is above all a prophecy of beauty-yet-to-come, yet-to-be-unveiled. We may, if we want, dispense with Soloviev's "supernatural future reality," and understand this as the prophetic vision of an other sort of life, seen in and through biological life, which remains always not quite here but nevertheless perceptible to those who study.

Beauty thus has the potential to pull us into a world forever being transfigured by those who study it. It can open up a future that is also the present remade and freed of the discipline imposed by the sublime in all its various forms. It is a world of endless depth and endless transformation, unbounded by fear or control. It is also inevitably a form of co-creation, as the studious senses participate in the creativity of nature, seeing as Soloviev says above, in "flashes" what is in any case always already there.

Study, then, is an encounter with nature that reveals a boundless future that is always approachable and knowable but never graspable. It is always more, always an excess pulling us onward. This encounter is also an experience that has long been known. In the words of Gregory of Nyssa:

> We can conceive of no limitation in an infinite nature; and that which is limitless cannot by its nature be understood. And so every desire for the Beautiful which draws us on in this ascent is intensified by the soul's very progress towards it ... [N]o limitation can be put upon the beautiful, and ... the increase in our desire for the beautiful cannot be stopped by any sense of satisfaction.
>
> (1979, pp. 147–148)

Our desire, released from the shackles of the sublime, turns to the beautiful; and since what we desire and what we desire to learn shapes us, at the same time we become aware of our own infinite depths, stretching out into a future form suspected but never reached. Study is always a knowing that is a kind of unknowing: we never know what the studious gaze will reveal in us and in nature, but we do know it will be even better than what is revealed in any given moment. It is dynamic and never-ending. It is free of knowable outcomes. Instead, it releases an instinct that life and liveliness taste of eternity, unlike "this dead and animal life" to which we are accustomed (Gregory of Nyssa, 2012, p. 371).

This understanding moves well beyond sustainability understood as "*sustaining* life or systems or institutions or persons or technologies or cultures or learning or knowledge or even the natural environment" (Barnett, 2018, p. 42). It moves, too, beyond Barnett's view that the ecological university should "be concerned with advancing or strengthening or positively *developing* life in all its forms" (2018, p. 42). Barnett's ideas still smack of an investment in control, in planning the future. The student of beauty, by contrast, abandons control for unpredictable transformation.

The managerial sublime or any attempt to deny students this experience is an attempt to bar them from these possibilities, to clamp down on life and blot out the future. It has the stench of death about it.

It is also worth pointing out that the infinite increase and never-ending transformation to which beauty frees the person studying is not based on a fantasy of economic expansion or a procession of products that claim to be "enhanced" or "updated." It does not rely on consuming ever more and ever newer objects. It does not require that we learn ever more IT applications just to keep up or worse, to be able to stay where we are. It does not expect us to live with the constant strain of re-organisation and re-structuring at work. All those just use up more resources and energy (and time). Beauty opens and addresses an ever-expanding and changing and never predictable future which fills but never fills by orienting our ecstatic desire.

If study opens up the human person to beauty then, equally, if it has not already been disciplined into oblivion, beauty can attract us to study in the first place. "Beauty 'bids' all things to itself (whence it is called 'beauty') and gathers everything into itself" says the fifth- or sixth-century theologian known as Dionysius the Areopagite (Pseudo-Dionysius, 1987, p. 76), making what is a traditional pun on *kalos* (beauty or goodness) and the verb *kaleo*, "to call, or bid or summon" (see Chryssavgis, 2019, p. 95). Beauty calls all to it, if we are not prevented by fear, and its call is *eros*, which is yearning or desire. Gregory of Nyssa's *Homilies on the Song of Songs* constitute a lengthy meditation on the awakening of eros by beauty. As the Bride is able to gaze ever more closely on the Bridegroom, she "sees the unspeakable beauty of the Bridegroom with a pure eye and in this way is wounded by the incorporeal and fiery arrow of love, for *agapē* when intensified is called love [i.e., eros]" (Gregory of Nyssa, 2012, p. 403). "[S]ome of our writers on sacred matters," argues Dionysius the Areopagite, "have thought the title 'yearning' [i.e., eros] to be more divine than 'love' [agape]" (Pseudo-Dionysius, 1987, p. 81). This desire, this yearning, this erotic calling is the energy with which beauty draws us endlessly towards it.

Study is nothing but a commensurate response to this erotic calling of an infinitely beautiful world. It answers with its own powerful desire or yearning, its devotion or pleasure, its energetic eros. Hearing and following the call of the beautiful, we pursue our vocation – our personal calling – and as such is in stark contrast to the ideology of employability. Employability disciplines students and teachers into conforming to the given requirements of employers. As those given requirements are known in advance, employability merely manufactures the same rather than an unpredictable and ecstatic future of beautiful and infinite hyperplenitude. It too has the stench of death about it.

Our vocation, then, is ecstasy, as we are drawn on forever into the infinite future of the world, and the world is what it is and becomes what it is and will be endlessly. Not that the world around us disappears – rather we see in and through it ever more those flashes of beauty mentioned by Soloviev. As we study, we develop what Ware calls "'double vision' … [B]y a strange paradox the more a thing becomes transparent to endless depth, the more it is seen as uniquely itself" (2013, pp. 94–95). Ware explains, "We are to perceive each kingfisher, each frog, each human face, each blade of grass in its uniqueness. Each is to be real for us, each is to be immediate" (2013, p. 94). Only when to the eyes of the student is revealed the utter irrefragable presence of a particular tree, are they able to perceive the infinity of beauty within it. Ware makes clear (2013, p. 94) that he is referring to William Blake's statement, in a letter to Thomas Butts, that "a double vision is always with me" (Blake, 1979, p. 461). Study's yearning gaze is essentially poetic and imaginative and abjures the perceptions of the scientific literalist for being in essence comatose: Blake concludes, "May God us keep/From Single vision & Newton's sleep!" (1979, p. 462).

All human beings should study to become poets, to hand back the future to the environment and also of course to themselves. We and all beings with us deserve our fathomlessly free beauty.

The word "poet" is not meant to imply that we should all begin scribbling verse, however. Why should one do that if it is not one's vocation? Poetry is, rather, the quality of vision one brings to whatever one's calling is: loving, yearning, desirous of ever more, knowing too that that excess is always about to be. All subjects should be a kind of applied poetics, as should all professions (no matter how much the sublimely masculine may sneer).

How may study achieve this? Through attention to beauty, of course. But that is not enough. Perhaps education should also awaken a sense of *something else*, a sense of *otherliness* in the world around us. Not an uncanny presence, but a depth in which we can find our own depth by being in relationship to it. An Other which is beyond all our attempts to conquer, control or define, but for which we may yearn, and which opens up the possibility of ourselves becoming freed from sameness or fear, to be also utterly transfigured and transformed. Something which restores to the world and ourselves a future, here and now and always. This is a question of a style of teaching which both educates and evokes, in the etymological sense of calling something out of the student, something which cannot be known in advance. Of course such a view implies that teachers are themselves calling forth something which they do not and cannot control, something which is other from the teachers. Perhaps this is what Gert Biesta, following Levinas and Kierkegaard, calls "revelation" (2016, pp. 50–51; 2017, pp. 52–54), or "something that comes from the *outside* and brings something *radically new*" (2016, p. 52, italics in original). Evocation, in other words, requires that the something else calls though the teacher, but in exactly the way that that particular teacher would put it. In evoking, the teacher is giving voice to their vocation, which, by its very nature is always present and always beyond them and ever poetic.

My own teaching

In my own teaching, I prefer to focus on somehow letting in the infinite beauty of the Something Else. This may be facilitated through the content or its presentation. I have described elsewhere (Wilson, 2020) my practice of approaching a specific landscape through its portrayal in folklore, literature, poetry, history, and archaeology. These narratives transform the territory into a mythopoeic land, which is then walked with students. It is hoped that they may be able to see it with new eyes primed for its numinous poetry, to hear its call and be drawn into it, so that "love calls to love" (Wilson, 2020, p. 135). If nothing else, my students report back on a fascinating experience.

As I learnt myself in my local countryside, walking is an excellent way of opening the senses. Even talking the walk, if one is unable to leave the classroom, may do the trick. Sessions in several different modules have been dedicated, for example, to Joseph Ferdinand Cheval – the so-called *Facteur Cheval* – obscure French postman, prolific walker, and builder of the extraordinary, the visionary edifice *Palais Idéal* (see Cardinal, 1972, pp. 146–153; Wilson, 2012). This is the story of how a sudden and unexpected encounter with an infinitely creative power in and through nature called forth a commensurate creativity in an ordinary man, revealing to him in a flash his vocation.

The Facteur became a postman in 1867, aged 30. His round covered some 20 miles (32 kilometres), and he did it on foot in all weather, over ground which was frequently rough, carrying a heavy satchel. To beguile the time, he explored in his thoughts a huge castle of his own invention, extending it year by year. In an autobiographical *cahier* of 1911, he wrote this:

> [C]onstantly walking in the same surroundings, what could I do but dream? What about? ... Well! As a diversion for my mind I built a fairy-like palace beyond imagination with all that the genius of a humble man could conceive (with grottoes, towers, gardens, castles, art galleries and sculptures) trying to bring back to life all the ancient architectures of primeval times; the whole thing was so pretty, so picturesque that for at least 10 years I could remember it vividly in my mind.
>
> (Cited in Jouve, Prévost and Prévost, 1981, p. 298 [author's translation])

After ten years, however, the fairy palace of imagination vanished. He found himself possessed by a powerful yearning, a profound desire to wander the castle in his body as well as in his mind. But this very desire blocked his ability to conjure it, as he realised that he did not have the skills to construct it in the world outside his mind. Evicted from the place of his longings, he trudged his round for a couple of years, until a sudden jarring shock to his body opened him up to what seemed to be an infinite creative power within nature itself. In April 1879 he stumbled and almost fell over something. Turning, he saw a rock resembling "a sculpture as bizarre as it is impossible for man to imitate, it represents every species of animal, every kind of caricature. I said to myself: since nature wants to sculpt, then I will turn my hand to masonry and architecture" (letter by the Facteur Cheval, 15 March, 1905, cited in Jouve, Prévost and Prévost, 1981, p. 298 [author's translation]).

As experienced by the Facteur, an infinite creative willpower possessed by nature itself breaks in, releasing his own power to make. He is unstoppably and unpredictably transformed as he builds the astounding masterpiece of visionary architecture called the *Palais Idéal*. On one of its facades, he writes his name in large letters: CHEVAL (Jouve, Prévost and Prévost, 1981, p. 286). The Palais *is* his transfigured self, immortal as stone, fascinating, and, above all, endlessly beautiful.

The point of the session, taught in modules on spirituality, imagination, landscape or even ethics, each time slightly altered, is to evoke the transformative effect of an inner desire or yearning meeting, and recognising, an infinitely creative and beautiful power in and through nature. As the Facteur builds he finally comes to life: he commences his work with a feature called "the Fountain of Life," in acknowledgement that the poetic power irrepressibly flowing through him is life itself. He is employed as a postman, but his vocation is to create: as he writes around one of the Palais's features, "*La vie sans but est une chimère*" (Jouve, Prévost and Prévost, 1981, p. 287): Life without a purpose is an illusion.

This is a story of an artistic – a "poetic" – relation to the natural world. The point of narrating it is to evoke and encourage a relationship between human persons and nature that is more than sustainable: it is a mutual intertwining of creative yearning that reveals infinite possibilities on both sides. All are carried forward and transformed in this deep connexion. I tell the tale, illustrating it with photographs of the Facteur and his family, acting out elements including his momentous stumble. Each time, the students have become intrigued and identify with this strange man (the Facteur, not me). At a given moment I judge propitious I show them an image of the Palais: without fail, there is a general gasp, and the students are transported. Whether any definite transformation is called forth I cannot say: but surely a subtle inner rearrangement occurs.

Somewhere between the more extensive walks and the classroom-based sessions lie brief but strange guided walks on campus, which aim at changing the vision of the participants. My fellow walkers proceed in pairs, one with eyes shut, the other guiding. The walk is slow, meditative, labyrinthine, and punctuated by my reading aloud short texts of my own composition. One such mazy stroll, for example, is based on the classic film, *A Canterbury Tale* (1944). The first text is in fact a description of a walk undertaken by one of the film's characters:

> Towards the end of the film *A Canterbury Tale*, written, directed and produced by Michael Powell and Emeric Pressburger in 1944, Alison Smith, played by Sheila Sim, wanders the bombed-out streets of Canterbury, which have been reduced to rubble – flattened – by the blitz of June 1942.

> She is looking for a particular garage, but in a cityscape devoid of the usual points of reference, she loses orientation completely.

> She asks a passer-by the way and is told, "It is an awful mess: I don't blame you for not knowing where you are. But you get a good view of the cathedral now" (*A Canterbury Tale*, 1944).

> Alison turns around and sees, as if for the first time, the cathedral. It is as if it had simply not been there before and is now revealed by the absence of buildings around it.

With the aid of this sight, she is now able to orient herself, and eventually find the garage, where she will be told that her lover, whom she believed killed in action, has in fact survived, and is returning to her. She is overcome by joy.

Alison and the city of Canterbury are both suddenly and unpredictably transformed by the unexpected appearance of the cathedral, which thus plays a similar role to the Facteur's stumbling block. The beauty of the ancient church, hitherto invisible to her, appearing in and through a world literally flattened and disfigured by the actions of man, leads Alison out of despair into fulfilment, drawing her ever forward for what will be the rest of her life. She is the Bride of the *Song of Songs*, to whose astonished eyes the Bridegroom is revealed, and an eternity of ever greater *eros* is opened:

> As, then, the veil of hopelessness is lifted and she sees the infinite and unlimited beauty of her Beloved, a beauty that for all eternity of the ages is ever and again discovered to be greater, she is pulled by a yet more intense yearning.
>
> (Gregory of Nyssa, 2012, p. 389)

It is hoped that this walk may change the relationship of participants to the campus and thereby the University, which may thereby, at least for a moment, be freed of the dead hand of the managerial sublime and become a place of study, that is, life.

Of course, I can never know what the effect on students of the walks and talks will be, if anything. Nothing is guaranteed. But there is still the mad wager that students, like the Facteur Cheval, will fall headlong in love with their studies and with the world, to plunge without fear, and without restraint, ever further into a mutual embrace that will never end and which is always uncontrollably in excess of what it is: delight.

Notes

1 "Eros," here, does not primarily imply sexual desire, but energy and delight-filled, dynamic, ever-changing, gratuitously excessive and unpredictable relationships, alive and life-giving.
2 See too Tansy Watts in this collection, who, following Thomashow (2002), contrasts enlivening qualities of attention to, and study of the natural world with more conventional forms which, in their refusal of relationship, are alienating and deadening.

References

Barnett, R. (2018) *The Ecological University: A Feasible Utopia*. Abingdon: Routledge.
Basil the Great (1978) "The Hexaemeron," in Schaff, P. and Wace, H. (eds.), *A Select Library of Nicene and Post-Nicene Fathers of the Christian Church: vol. 8 St*

Basil: Letters and Select Works. 2nd Series. Grand Rapids: William B. Eerdmans, pp. 51–107.

Biesta, G.J.J. (2016) *The Beautiful Risk of Education.* Abingdon: Routledge.

Blake, W. (1979) *Blake's Poetry and Designs.* Edited by Mary Lynn Johnson and John E. Grant. New York: W.W. Norton.

Burke, E. (1998) *A Philosophical Enquiry into the Origin of our Ideas of the Sublime and Beautiful and Other Pre-Revolutionary Writings.* Edited by David Womersley. London: Penguin Books.

A Canterbury Tale (1944) Directed by M. Powell and E. Pressburger [DVD]. Reissued, UK: Granada Ventures (ITV DVD) Ltd, 2007.

Cardinal, R. (1972) *Outsider Art.* London: Studio Vista.

Carnes, N. (2014) *Beauty: A Theological Engagement with Gregory of Nyssa.* Eugene: Cascade Books.

Chryssavgis, J. (2019) *Creation as Sacrament: Reflections on Ecology and Spirituality.* London: T&T Clark.

Chryssavgis, J. and Foltz, B.V. (2013) "Introduction. 'The Sweetness of Heaven Overflows onto the Earth': Orthodox Christianity and Environmental Thought," in Chryssavgis, J. and Foltz, B.V. (eds.), *Toward an Ecology of Transfiguration: Orthodox Christian Perspectives on Environment, Nature, and Creation.* New York: Fordham University Press, pp. 1–6.

Eagleton, T. (1990) *The Ideology of the Aesthetic.* Oxford: Basil Blackwell.

Foltz, B.V. (2014) *The Noetics of Nature: Environmental Philosophy and the Holy Beauty of the Visible.* New York: Fordham University Press.

Gregory of Nyssa (1979) *From Glory to Glory: Texts from Gregory of Nyssa's Mystical Writings.* Translated and edited by H. Musurillo. Yonkers, NY: St Vladimir's Seminary Press.

Gregory of Nyssa (2012) *Homilies on the Song of Songs.* Translated by R.A. Norris. Atlanta: Society of Biblical Literature.

Hall, R. (2021) *The Hopeless University: Intellectual Work at the End of the End of History.* Mayfly Books.

Hart, D.B. (2003) *The Beauty of the Infinite: The Aesthetics of Christian Truth.* Grand Rapids: William B. Eerdmans.

Johnson, J. (2020) *The Father of Lights: A Theology of Beauty.* Grand Rapids: Baker Academic.

Jouve, J-P, Prévost, C. and Prévost, C. (1981) *Le Palais Idéal du Facteur Cheval: Quand le Songe Devient la Réalité.* Hédouville: A.R.I.E. Éditions.

Larchet, J-C. (2017) *Theology of the Body.* Translated by M. Donley. Yonkers: St Vladimir's Seminary Press.

Lea, D.R. (2011) "The Managerial University and the Decline of Modern Thought," *Educational Philosophy and Theory: Incorporating ACCESS* 43 (8), pp. 816–837.

Otto, R. (1958) *The Idea of the Holy: An Enquiry into the Non-Rational Factor in the Idea of the Divine and its Relation to the Rational.* Translated by J.W. Harvey. Oxford: Oxford University Press.

The Oxford English Dictionary (1989) Vol. 16 Soot-Styx (1989) 2nd edn. Oxford: Oxford University Press.

Pseudo-Dionysius (1987) *The Complete Works.* Translated by C. Luibheid and P. Rorem. Mahwah, NJ: Paulist Press.

Sammon, B.T. (2013) *The God Who is Beauty: Beauty as a Divine Name in Thomas Aquinas and Dionysius the Areopagite.* Eugene, OR: Pickwick Publications.

Sammon, B.T. (2017) *Called to Attraction: An Introduction to the Theology of Beauty.* Eugene: Cascade Books.

Smith, J.K.A. (2009) *Desiring the Kingdom: Worship, Worldview, and Cultural Formation.* Grand Rapids: Baker Academic.

Soloviev, V. (2020) *The Heart of Reality: Essays on Beauty, Love, and Ethics.* Translated and edited by V. Wozniuk. Notre Dame: University of Notre Dame Press.

Thomashow, M. (2002) *Bringing the Biosphere Home: Learning to Perceive Global Environmental Change.* Cambridge, MA: MIT Press.

Von Balthasar, H.U. (1982–1989) *The Glory of the Lord.* Translated by E. Leiva-Merikakis, A. Louth, F.McDonagh, B. McNeil, J. Saward, M. Simon, R. Williams and O. Davies. 7 vols. San Francisco: Ignatius Press.

Ware, K. (2013) "Through Creation to the Creator," in Chryssavgis, J. and Foltz, B.V. (eds.), *Toward an Ecology of Transfiguration: Orthodox Christian Perspectives on Environment, Nature, and Creation.* New York: Fordham University Press, pp. 86–105.

White, L. (1967) "The Historical Roots of Our Ecologic Crisis," *Science* New Series 155 (3767), pp. 1203–1207.

Wilson, S. (2012) "Fortean Traveller 79. The Palais Idéal," *Fortean Times* 286, pp. 74–76.

Wilson, S. (2020) "Walking into the Heart of the Landscape to Find the Landscape of the Heart," in Armon, J., Scoffham, S. and Armon, C. (eds.), *Prioritizing Sustainability Education: A Comprehensive Approach.* Abingdon: Routledge, pp. 126–140.

Wilson, S. (2022) "The Fierce Urgency of Now and Forever and unto Ages of Ages: Study and the Restoration of Paradise on Earth," *International Journal of Sustainability in Higher Education* 23 (1), pp. 29–40.

Zizioulas, J. (2021) *Priests of Creation: John Zizioulas on Discerning an Ecological Ethos.* Edited by John Chryssavgis and Nikolaos Asproulis. London: T&T Clark.

4 Educating for the future in the humanities

Passion, utility and student perspectives of employability in higher education

Claire Bartram and Chiara Hewer

Introduction

This chapter asks how the toxic ubiquity of the economic metric currently used to measure the value of higher education can be counteracted. It reaches into wider debates about the value of the humanities as a discipline, and asks how students can be enabled to consciously value the skills for employment that their degrees hone and recognise, 'how well their education sets them off into the world of work' (Department of Education, 2022 p. 21). In the face of a government-driven economic narrative that promotes the financial returns for the individual, the taxpayer, and the exchequer, passion for a subject, the typical motivator for choosing a humanities degree, is devalued as a false rationale. A powerful conclusion of this instrumentalist narrative is that students want 'value for money from their own investment', and degrees with a clear vocational outcome, particularly in medicine, economics, and law are the best means of realising this value (Department of Education, 2022 p. 21). The pursuit of a passion becomes an irrational and financially irresponsible action. As Joe Moran summarises, in the language of economics, 'University education is a disutility – the sacrifice of one's time and convenience for money' for which the only valued outcome is a salary that justifies the debt (Moran, 2022, p. 20).

The humanities are complexly situated in relation to this 'hyper-utilitarian' (Morris, 2017) narrative. Helen Small's case study of Matthew Arnold (Small, 2013 pp. 81–85) illustrates long-held resistance to instrumentalism, seen in Arnold's relentless editing out of the word 'use' and his refusal to embrace Platonic associations of utility in his translation of '*Dulce et Utile*' as the 'sweetness and *light*'. Small (2013 p. 175) established a 'pluralistic defence' of the humanities examining key arguments for public value. Whilst reminding us that 'any defence that gives primary place to the instrumental value of a humanities education will quickly disfigure the broader kinds of good it nurtures', Small also suggested that utility 'no longer figures quite so plainly as the enemy' (2013, pp. 175, 67). More recently, Zoe Bulaitis (2020, p 15) has extended this discussion of value, counselling against a reactive response to 'neoliberal trappings of singular answers, irrefutable data, and tangible

DOI: 10.4324/9781003286516-6

results'. Bulaitis asserts that recognising the longevity of debates and recurring crises for the humanities across the last hundred-plus years enables a more measured response that seeks to articulate the multifaceted work and value of the humanities. She reminds us that 'the word "value" should not simply concern economic value, but also social, ethical, and moral values. Neoliberalism would have us forget that economic value is just one voice among many'.

The chapter contributes to current debates by examining the passion-utility dichotomy in a precise political context of the Johnsonian government (2019–2022), exploring counter-narratives, and tracking the British Academy's stance on the value of humanities education. Against this national backdrop, it juxtaposes an institutional response to the utility narrative by focusing on the employability agenda for the humanities at Canterbury Christ Church University. Drawing on student experiences of employability in a Level 5 module *Applied Humanities: Employability in Practice*, the chapter considers whether this instrumentalist imposition on the curriculum can function as a rhetorical means of aligning these oppositional narratives of passion and utility.

Value for money, the humanities and the rise of the employability agenda

The issue of student finance has been a key driver in the development of an economically defined, marketized approach to measuring the value of courses and subject areas. The value-for-money narrative was significantly heightened in the Johnson administration and has been perceived as anti-humanities in stance in its promotion of STEM (Science, Technology, Engineering, Maths) and skills-led or vocational courses. The newly available Longitudinal Educational Outcomes Survey (LEO) data together with an annual student loan rate of close to £20 billion a year and an expectation that only about a quarter of the students will repay their loan in full, has intensified scrutiny of higher education. Recent government reports emphasise that higher education is an 'investment' that must provide 'good value for money' and 'good outcomes' based on the commodities of 'skills and knowledge'; and in that order (Department for Education, 2022 p. 31).

In the value-for-money narrative, humanities degrees are presented as performing poorly. By correlating student loan information with tax records, the LEO identifies high and low financial returns of degree subjects measuring success in terms of income and repayment of student finance. The Institute for Fiscal Studies (IFS) report on projected lifetime earnings reinforces the fiscal imperatives of good value and good outcomes as directly related to high incomes, which the fields of medicine, economics, and law facilitated, but degrees in the creative arts and English literature apparently did not (Department for Education, 2020, p. 6). As Andrew McGettigan (2015) foresaw, 'universities and colleges are then to be judged on how well they provide training that does indeed boost earnings profiles'. In this marketized 'anti-vision of university education', McGettigan (2015) highlighted a potential Treasury move to risk-assess the likely repayment of student loans by graduates of specific universities. The value for money of a humanities degree is not only measured in terms of graduate salary outcomes but also in the cost of delivering a course. As Gavan Conlon noted, 'there was no incentive for universities to charge anything other than [the full] £9,000', for 'institutions charging less could be considered to be offering *lower quality* qualifications' (Hubble and Bolton, 2018 p. 9). The introduction of student fees failed to create a competitive market, and no cost differentiation was made by institutions for the provision of courses that required expensive clinical, laboratory, studio, or fieldwork learning environments alongside courses that did not. The Augar Report speculated whether tuition fees should vary by subject and that perhaps government would move to cap student numbers for courses perceived to be most burdensome to the taxpayer in terms of unrepaid student loans (McVitty, 2019).

The value-for-money narrative was further compounded by Johnsonian education ministers, who caused consternation in their characterisation of students as victims of some kind of educational fraud. Michelle Donelan, a recent Universities Minister described how 'those for whom university is a new and unfamiliar world are most vulnerable to falling onto courses where their graduate prospects are poor'. Her emphasis on the naïve vulnerability of the non-traditional student duped by 'poor-quality, rip-off courses' (Donelan, 2022), mirrors comments made by her predecessor, who described an environment that was 'pushing young people on to dead-end courses that give them nothing but a mountain of debt' (Hazell, 2021). Gavin Williamson, speaking here, emphasised the need for '*valuable* and technical courses' that addressed the 'gaps in our labour market' and are what 'our society *needs*' (Williamson, 2021, author emphasis). Funding for Arts subjects was cut three months later.

The rise of the employability agenda is an inseparable part of this instrumentalist position in which higher education is valued as 'a human capital investment that *benefits* the private individual insofar as it enables that individual to boost future earnings' (McGettigan, 2015). Paul Blackmore and Camille Kandiko (2012) point to the commodification of the university experience, noting that globally universities have 'increasingly assumed a responsibility for promoting economic growth, especially by providing "job–ready" graduates'. Two notable reports by the Department for Business, Innovation and Skills in 2011 were instrumental in foregrounding employability as a priority. The White Paper *Students at the Heart of the System* re-evaluated the perceived value of higher education and its qualitative benefits while the Wilson Review of *Business–University Collaboration* identified the need for stronger connections between universities and employers. In response, universities were pressurised into establishing more tangible relationships with the private sector to meet demands for a more prepared workforce, were subjected to additional quality-control systems, and were told to come to terms with the idea that students were 'customers', particularly following the introduction of higher university fees, and had the right to expect a 'return' on their significant investment in education. In addition, metrics from the Destination of

Leavers from Higher Education (DLHE) survey, established in 2003 to report on employment destinations of university graduates six months after graduation, increasingly drove higher education institutions' employability agendas particularly when linked by Government policies to student funding and used as performance indicators in university league tables.

At Canterbury Christ Church University, employability became one of the eight cross-cutting themes in the institutional *Strategic Framework* (2015–2022). Identified as one of the University's key performance indicators, the University set an institutional target of achieving 65 per cent in graduate-level employment, and 94 per cent in employment and/or further study, by 2020. The University's strategic framework set out a distinctive and broad student experience with opportunities for external engagement through placements, internships, study abroad, language-learning, and community engagement as part of developing intelligent citizenship. It committed to developing a distinctive curriculum experience for all students through embedding employability, internationalisation, and sustainability. Whilst the last DLHE (2017–2018) figure, before the introduction of Graduate Outcomes, showed graduate-level employment at six months had improved for the fourth year running (at that point in time 63.1%) and was on track to achieving the institutional target, the University's performance remained significantly below the sector average (73.6%). In addition, the University's Employment Indicator figure fell for the first time since 2012/13 (to 93%) and was below the benchmark. With Graduate Outcomes set to be a key metric in the Teaching Excellence Framework (TEF), the need to improve student progression into highly skilled (graduate) employment remained a priority across the university as set out in its Employability Framework *Future 360* (2019–2024). The framework is informed by Advance HE's view of employability as providing 'opportunities to develop the knowledge, skills, experiences, behaviours, attributes, achievements and attitudes that enable graduates to make successful transitions benefitting them, the economy, and their communities' (Tibby and Norton, 2020, p. 5).

Counter-narratives of value

As Eleanora Belfiore notes, econocratic approaches to policymaking set cost-benefit analysis at the heart of the decision-making processes producing an 'ideologically disingenuous' and 'essentially defensive recourse to narrowly utilitarian rationales' (Belfiore, 2015 p. 98, 105). Belfiore draws on Brewer's lament at the resulting 'confusion over what value means and the narrow way in which it is often employed' (Brewer, 2013, cited in Belfiore, 2015 p.105). This is evident in recent government reports in which the IFS (Department for Education, 2020 p. 70) for instance, is unable to use the word 'value' for non-economic outcomes, reminding us that it does not consider the 'non-financial *benefits* to higher education such as improved health or happiness, nor any *wider returns* to society' in its analysis. Notions of 'social' and 'cultural' value also flit past with the Augar Report noting the '*social value* of some lower-earning professions such as nursing and social care, and the *cultural value* of studying the Arts and Humanities' (Augar, 2019 p. 91, author emphasis). While it asserts that 'we are clear that *successful outcomes* for both students and society are about more than pay', Augar cannot define or quantify these with the effect that the economic instrumentalist argument has a disproportionate influence.

If we are searching for other indices of value, Bulaitis cautions that we will not find them in this 'myopic neoliberal narrative' and must look beyond 'the terms of debate defined by economic white papers and policy documents' (2020 p. 27, 29). Counter-narrative strategies often involve a desire to resist the language of 'the bureaucratic statistician' (Small, 2013, p. 87), and one striking approach is seen in a life-affirming counter-narrative of love. This is evident in the emotive conclusion of an Arts and Humanities Research Council report that 'understanding history and culture … makes life worth living'. The statement uses interestingly positive, nurturing vocabulary, it '*supports* learning, *stimulates* innovation and creativity, *enhances* the potential of the UK's growing and uniquely productive creative economy' (2013 p. 24, author emphasis). There are other similar rhetorical matrices in use. A British Academy (2021) press release talked about the need to ensure the 'health' and 'vitality' of university degree provision, of 'growing' and 'thriving', and of the dangers of regional 'cold spots' in HE provision when degree courses are closed. Hetan Shah, chief executive of the British Academy, has encouraged humanities students to 'feel reassured they can study what they love and have a great career at the end of it' in a rhetoric that values passion and locates it within a broader institutional narrative of health, growth, and warmth (Bennett, 2020). Similar language is seen in historian Robert Tombs' assertion (cited by Heffer, 2022) that 'the humanities are the core of our cultural existence. We need to cherish them'. This approach is captured in Zoe Bulaitis' recent call for 'an articulation of the value of the humanities [that] encompasses the lives, ideas, and values of people as opposed to their instrumental use as products' (2020, p. 3).

While the British Academy has been a vocal opponent to the utility narrative, arguing that, 'degrees are about more than employment and more than giving you skills for the economy', it has also developed a skills narrative for the humanities. A humanities degree is valuable, it suggests, precisely because it is not specifically vocational but provides 'a wide skill set, including communication skills, collaboration and teamwork, research and analysis, independence and creativity', enabling 'flexibility and resilience' in the workplace (Reidly, 2021). This narrative has been accompanied most recently by the foundation of SHAPE (Social Sciences, Humanities, and the Arts for People and the Economy /Environment). Described by Julia Black (2020), president of the British Academy, as comparable to STEM (Science, Technology, Engineering, and Maths), the initiative aims 'to "level up" the profile of SHAPE to be on a par with STEM, not in opposition but as equals, and as collaborators'. What we see in the founding of SHAPE is a promotion of parallel skillsets for an interdisciplinary future as advocated by Shah (2021): 'humanity's insight is more robust when STEM and SHAPE come together'.

There is also a strategy of scrutinising and in some instances rejecting the validity of the economic data. We see this in the assertion made by David Duff, chair of the English Association, that one simply cannot 'measure the value of education with a calculator' (Fazackerley, 2021). The British Academy (2021 p. 5) agrees: 'most importantly, using graduate earnings as part of a regulatory approach fails to recognise that the return from higher education cannot be measured solely, or even primarily, in economic terms on the basis of salary achieved'. This is reinforced by their assertion that a high salary is not a primary motivator for students and that when surveyed, they ranked interest in the subject and enjoying learning as higher priorities (British Academy, 2020, p. 15). It is most evident in Hetan Shah's uncharacteristically blunt assertion that a government-driven emphasis on STEM simply fails to reflect the economic reality: 'sometimes politicians can latch on to a particular narrative and it isn't always rooted, actually, in the data' (Baker, 2021).

The British Academy (2017, 2020) has also taken an evidence-based approach to examining the economic value of degrees in recent reports and has argued for a more nuanced analysis. It notes that 'much of the variance in graduate earnings is due to underlying characteristics such as socio-economic background, prior attainment, and institution attended rather than subject of study' (2020, p. 16). Most recently, it has refuted the use of the LEO data, questioning the narrow definition of 'graduate employment' as prejudicial to a significant range of employment sectors popular with humanities graduates including the creative industries. This sector is noted to be one of the 'most dynamic, productive and profitable' with fast growth predicted over the next decade, yet characterised by a high level of self-employment, it is under-represented in the IFS definition of a 'graduate' or 'managerial' role. (British Academy, 2020 p. 22) This demand for nuance is a forceful strategy that has the potential to undermine the surety of the assertion that the LEO data can 'provide *valuable* information to students making their choice of whether to go to university and what to study, but also provide policymakers with *important* information on the *value for money* of different degrees'. (Department for Education, 2020 p. 70, author emphasis)

David Willetts, instigator of the LEO project as minister for Universities and Science, himself highlighted the limitations of early LEO datasets. At one strategically poignant moment in a newspaper article, Willetts (2019) presents an example of 'a first-generation university graduate who works part-time in her local area as a nurse'. He notes that 'she does not show up well in the [LEO] data, but she might still have been on a significant personal journey, with wider social and economic benefits'. In pitting this fictional nurse against the data, Willett's story invites us to ask what we should value and how we capture the 'significant personal journey' invisible in the data. Willett's intervention here is interesting stylistically. It brings the human into the data, and it incites emotional and moral responses from the reader. This nurse matters, she is valued in her community, but she is at once an example of social mobility and simultaneously an emblem of a wider social reality in which her fractional employment status, her gender, and the low-earning potential of her profession mitigate against her significant personal achievements. It is interesting that as a politician, Willet interrupts the surety of the data-driven instrumentalist emphasis with a narrative reality check. It contrasts with much of the other imperative but ill-defined rhetoric about 'what our society needs'.

Ten years ago, Helen Small (2013, p. 10) asked whether it was possible that 'we now inhabit a public sphere so distortedly geared to thinking in terms of economic profitability that we need corrective input from the humanities to redirect our attention to human goods more variously described?' Arguably, the Johnsonian period of government saw a toxic escalation of the neoliberalist rhetoric and values that demanded that the 'importance of humanities has to be proven constantly'. What is particularly noticeable about this period is the development of a response that takes the form of 'a continual evidence-based rebuttal' (Baker, 2021). Whilst promoting the passion ethos, the British Academy also produced a skills narrative, provided data analysis, and was instrumental in launching a parallel brand for the humanities as a part of SHAPE, in a strategy that was both passion- and utility-driven. In the same period, Willets (2019) appears to provide an example of a move away from a utility narrative towards a rejection of neoliberalism as a meaningful unit of value embracing precisely the kind of approach asserted by Bulaitis, to provide a 'humanistic valuation' (2020, p. 7).

Valuing employability

Moving to examine instrumentalism at an institutional level, this section tracks the development of a bespoke employability module for humanities students at Canterbury Christ Church University. It explores how the value or learning gain of the module can be captured using an Abintegro Career Pulse questionnaire and couples the data generated by the questionnaire with the students' reflective responses to the questionnaire results. It then reviews the reflective assignments more broadly to explore the stories students tell about themselves, their experiences and expectations, and how they locate themselves in relation to the world of work pre and post-placement. As such, this section explores the interface between utility and passion drawing on institutional policy-making, data analysis and close reading.

Module design and learning gain

Applied Humanities: Employability in Practice was designed and validated in the academic year 2014–2015 as a joint project between the Career and Enterprise service and the then School of Humanities. The initial option considered by the Careers and Enterprise team was to design an 'off-the-shelf' generalist employability module which course teams could make available to students as an elective module. However, it was soon clear that this 'one size fits all' approach would not serve the needs of our diverse student population, or the industry requirements for the roles identified as likely career paths. Both demanded meaningful tailored provisions and not just another 'tick-box' exercise. Most colleagues seemed positive about the possibility of an employability module and recognised the opportunities that this would pose for developing students' skills as well as strengthening contact with alumni and the wider professional community. There was some resistance to the initiative from academics anxious that the module would not match the academic rigour of their wider programmes. Such concerns were addressed through the active engagement of the Employability Manager with student focus groups, alumni, local employers, and colleagues from academic teams and professional services to identify evidence-based research literature and meaningful learning outcomes. Launched the following academic year, the pilot suite included students from History, American Studies, English Literature, Theology and Religion, Philosophy and Ethics courses.

The module was inspired by two established graduate employability models: Knight and Yorke's USEM model (2004) and Dacre Pool and Sewell's CareerEDGE model (2007). Both 'recognised the needs of students, employers and other stakeholders … [to develop] the student's efficacy and metacognition and relate this to the development of subject knowledge and professional skills that are transferable to the practice context' (Cole and Tibby, 2013, p. 7). Feedback from employers confirmed that the ability to articulate how skills and qualities could be applied in a work context was the most important aspect of graduate employability. The aim of the module was, therefore, to enable students to identify the skills gained through their undergraduate study and reflect upon how they may be applied to work-related activities. In contrast to academic class-based learning, the focus was on practical work-based experiences such as a placement, industry-informed case study, or employer-led project work. Whilst such an approach was relatively common practice in other disciplines, the content and module structure was innovative in a humanities setting in the sector at the time.

Based on a pedagogical understanding that 'learning and doing cannot be separated and therefore to use knowledge at its fullest potential it must be implemented, performed, enhanced as part of a synergy' (Heyler and Lee, 2014, p. 352), the module sought to provide an accessible way to understand the importance of employability 'to those new to the subject, and to students and their parents as well as appealing to academics' (Cole and Tibby, 2013 p. 8). The module design had three key components.

1. *Preparation for employment*, which includes interaction with industry speakers and alumni, building commercial awareness, and an understanding of career pathways available to humanities graduates.

2. *Work-related activity* in the form of micro-placements, live briefs, or negotiated projects, including the creation of employability assets such as CVs.
3. *Application of the work-related learning*, including student reflections on their experience, the application of discipline knowledge, and the students' own development of personal skills.

Reflective assessment is key to the module's success. Following the model established by Boud, Keogh, and Walker (1985), students are encouraged to turn experience into learning through reflective assessment. Using Pebblepad, the students complete two reflective portfolios. The first, weighted at 40 per cent of the final mark, invites reflection on workshop experiences with industry experts and the placement application process from designing an academic CV to completing an interview with a placement provider. The second portfolio worth 60 per cent, maps the placement experience and includes a reflective placement log, broader reflections on the module as a whole, and the next steps in the employability journey. As part of this reflective process, the students also complete a careers questionnaire that helps them to track their learning journey across the module. The Abintegro Career Pulse is a self-assessment questionnaire with 10 questions linked to key employability skills as identified in the university's *Future 360 Framework*. Students complete this self-assessment at the beginning and end of the module and reflect on their results in their portfolios. By comparing the two Career Pulse questionnaires, it is possible to measure the learning gain made by students by the end of the module. The table (Figure 4.1) sets out the learning gain for the three cohorts studied.

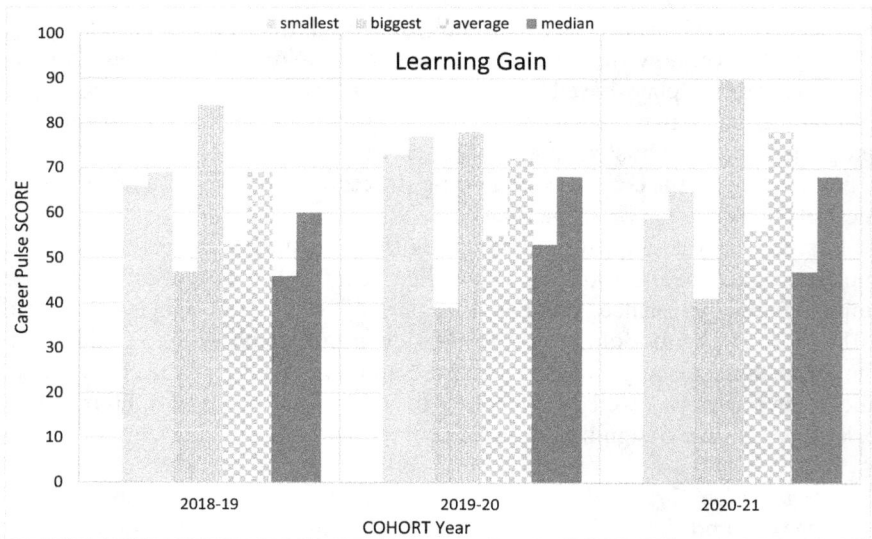

Figure 4.1 Career Pulse results for three cohorts showing learning gain.

By examining the reflective evaluation of the questionnaire in portfolios submitted by the students who scored the lowest, the median, and the highest learning gain, it is possible to identify learning gain in three key areas: the job application process, awareness of degree-skills, and enhanced self-knowledge. Most students recognised that they had attained a better understanding of the job application process with a noticeable increase in confidence regarding CV and cover-letter writing and interview skills. Many commented on the usefulness of combining best-practice sessions at the start of the module with the opportunity to put the learning into practice straight away as part of the work-related application process. They had attained a more in-depth knowledge of the sector and roles they wanted to move into after graduation and had a better understanding of how their degrees prepared them for their future careers. They also felt more confident regarding their ability to apply their degree subject knowledge and articulate its benefits to employers. Students also recognised that their employability development did not end with the module, but that action plans had to be put in place to continue working on areas of low confidence and develop new skills to give themselves the best opportunity to secure graduate employment. Many students noted an enhancement of their self-awareness and understanding of their skillset and acknowledged that the use of PebblePad, to record their learning and development throughout the module, had made them more reflective. Many showed an overall appreciation (and a somewhat heightened sense of pride) that most of the areas identified as 'areas for development' in the initial Career Pulse self-assessment, showed an improvement, though in various degrees, by the end of the work-related experience.

Student narratives of value

Reflective assessment and tutorial attendance invite students to evaluate their experiences and their performance and to engage with reflective modes of writing and thinking that continue to be a less familiar mode of assessment. In considering the student experience holistically in both formal and informal reflective settings, it is possible to identify authentic narratives. Student responses to the module are diverse and not always positive, but there are some broad trends that can be identified, and these are presented in three key narratives of belonging, resilience, and creativity. The approach in this section is narrative, often anecdotal, not data-driven and to protect confidentiality, student comments have been generalised or paraphrased.

Belonging

The module has a key role to play in enabling students to develop a narrative of belonging. For some students, this notion of belonging is linked to a particular career path that they want to pursue. Some students approach the module with a clear sense of vocation and through a placement experience related to their preferred sector, find affirmation of their plans. For those with less clear career plans, the placement experience can be a defining moment in discovering a sense of belonging to a particular sector: this can be an exciting revelation. Others benefit from discovering a broader sense of opportunity, that they could belong to a number of sectors. One commented that they were excited to consider the different career directions that they could pursue, and that the possibilities were far greater than they had initially thought. The practical placement also enabled a stronger sense of identity as a humanities student, to take ownership and be proud of a "marketable" skillset. One student spoke of feelings of enhanced self-worth that resulted from linking their passion for a humanities degree subject with a new-found career trajectory.

For some, the question of belonging is more complex and centres on a feeling of a lack of entitlement to belong to a graduate workforce. In the safe space of this module with hand-picked placement providers, students have an opportunity to confront some of their perceived barriers to employment. For some students, this is about requesting reasonable adjustments and managing their health and well-being in a workplace setting. For others, it is about aspiration shifting from low-pay and low-skilled work necessarily undertaken to aspire to graduate employment. For students with limited work experience, the issue of belonging is complicated by a perceived lack of experience and this concern can be shared by students returning to education. Within this group, caring roles are often initially discounted as being of little value by students. These students particularly benefit from the opportunity to actively draw together "life skills" with their degree skills to build a professional identity for the first time. For mature students, there is an opportunity to develop a new sense of belonging to re-enter employment in a new sector with a graduate skillset.

Resilience

Post-placement, students regularly commented on their growing confidence and sense of achievement in overcoming unexpected difficulties. Recurring observations of resilience include 'I never thought it would have been possible to', or 'I proved to myself my capability', or 'I achieved things I would have been far too nervous to do a few months ago'. There was also often a palpable sense of pride in achieving more than they expected, in discovering leadership qualities, and in recognising and articulating their success in the reflective work. Many of these students expressed a keen intention to continue to build their profile through further volunteering and some continued to work with their placement providers beyond the module. For others, the work placement experience encouraged them to seek paid employment by proactively joining or returning to a workforce that they previously felt excluded from. There is anecdotal evidence that the placement can be a transformative experience with students discovering an identity and a strong sense of belonging, what we might recognise as a sense of vocation and an ambition that reinvigorates their sense of purpose in their studies.

There is also a narrative of resilience that emerges from the discussion of aspects of the module that were 'a struggle' or 'not enjoyable'. Unexpected challenges provide valuable opportunities for reflection. For a multitude of reasons, placements do not always go exactly to plan, and life inevitably intervenes in a live project. 'Discomfort' is a recurring term within this kind of experience as students step outside their comfort zone to request clarification of instructions, tighter project perimeters, or to raise concerns about how a project is progressing. Often students were surprised by their tenacity and were ultimately empowered by having to take the initiative. One student commented that despite the stress and frustration felt, they would continue to seek to push themselves to engage with the 'uncomfortable' in other contexts to 'thrive'. Students also did not shy away from acknowledging where their performance faltered. Recurring issues include time management and organisational skills. Some students recognised and addressed these challenges during the placement. For others, these challenges surfaced in the post-placement evaluation with one student concluding that they 'could and should have made more effort'. It is perhaps testament to the support of the module tutors and the exceptional attention of the placement providers that students are able to make these observations about motivation and many students concluded, 'I learned more about myself than I expected'.

Creativity

Working creatively was also a recurring theme. Many students really thrived in an environment in which they had a measure of autonomy and valued the opportunity to organise their time, confidently working independently and liaising with a team. Some students expressly valued the opportunity to travel or be outdoors and others noted their reluctance to work in an office environment in the future. 'Flexibility' and 'freedom' were recurring values as was the opportunity to present research in different formats and apply their written and spoken communication skills in different settings. Some students spoke of the module as a refreshing experience, providing some 'much-needed variety' with the placement offering a welcome 'change of pace'. Interestingly, a few students spoke of the recuperative value of the placement as a respite from academic burnout. One student described how it had enabled an impassioned and proactive re-engagement with the purpose of their studies, which had otherwise come to feel like a laborious means to an end.

Applied Humanities: Employability in Practice was sector-leading in its inception. It was designed by and continues to be delivered by a team who reflect a wider institutional ethos of exceptional commitment to the student experience, a genuine interest in the individual, and who share a holistic and sustainable vision of teaching. We want our students to thrive both in and beyond the classroom, but Kate Daubney's assertion that 'young people lack the language, concepts, and understanding to make the connection between what they learn and how it enables their future work and study choices' (2021, p. 103), mirrors my experience as module convenor for *Applied Humanities*. It seems more important than ever that we equip our students to have 'the self-awareness to identify, articulate, and put into practice the employability value that makes their goals achievable' (Daubney, 2021, p. 103). Ceding ground to the economic instrumentalist argument by introducing employability into the humanities curriculum has empowered students to take ownership of the passion narrative and articulate why they study the humanities and what skills they offer employers.

Conclusion

Written in what felt like an intensified period of crisis within the Johnson administration – of heightened rhetoric, a high ministerial turnover, and an attendant media-maelstrom – this chapter is unavoidably defensive of the humanities. In a challenging climate of declining recruitment, course closure, and a cost-of-living crisis, post-1992 institutions are both essential to supporting non-traditional students and most at risk from falling student numbers. This sector has a particular duty to continue to enable a broad and sustainable curriculum. Thirty years ago, Bill Readings warned of the dangers of 'the analogy that makes the University into a bureaucratic apparatus for the production, distribution and consumption of knowledge' and that locates students as 'consumers' rather than 'thinkers'. His work anticipated many of the current sector challenges, and he urged universities to develop 'an accountability that exceeds the logic of accounting' (Readings, 1994 p. 27, 163–164). As Bulaitis reminds us, 'The humanities should be at the forefront of efforts to remember, to revise, and to reform narratives of value' (2020, p. 247). Located within an intensified political context in which utility has become part of a strategic defence of the humanities, this chapter has sought to juxtapose the macro and the micro and to identify current life-enhancing, passion-driven counter-narratives to economic instrumentalism. In showcasing the very positive student experiences of the humanities employability agenda at Canterbury Christ Church University, it has also sought to provide an intervention in but not a resolution of, the passion–utility dichotomy.

References

Arts and Humanities Research Council (2013) *The Human World, The Arts and Humanities in our Time AHRC Strategy 2013–18*. Available at: https://ahrc.ukri.org/documents/publications/the-human-world-the-arts-and-humanities-in-our-times-ahrc-strategy-2013-2018/ (Accessed 06/07/21).
Augar, William (2019) *Independent Panel Report to the Review of Post–18 Education and Funding* Available at: https://assets.publishing.service.gov.uk/government/uploads/system/uploads/attachment_data/file/805127/Review_of_post_18_education_and_funding.pdf (Accessed 05/01/2022).

Baker, Simon (2021) 'Is STEM Growth Really Stunting the Humanities? *The Times Higher Education Supplement*, Thursday 19 August, 2021. Available at: www.times highereducation.com/features/stem-growth-really-stunting-humanities (Accessed:14/07/22).

Belfiore, Eleonora (2015) 'Impact', 'Value' and 'Bad Economics': Making Sense of the Problem of Value in the Arts and Humanities' *Arts and Humanities in Higher Education* 14 (1), pp. 95–110.

Bennett, Rosemary (2020) 'Employers Prefer Arts Graduates to Scientists' *The Times* 07 May 2020 Available at: www.thetimes.co.uk/article/employers-prefer-humanities-graduates-to-stem-degrees-2gnz9hhvf (Accessed: 14/07/22).

Black, Julia (2020) 'Shape – A Focus on the Human World' *Social Science Space* Available at: www.socialsciencespace.com/2020/11/shape-a-focus-on-the-human-world/ (Accessed: 14/07/22).

Blackmore, Paul and Kandiko, Camille (2012) *Strategic Curriculum Change: Global Trends in Universities* Oxford: Routledge.

Boud, David, Keogh, Rosemary and Walker, David (1985) *Reflection: Turning Experience into Learning* Oxford: Routledge.

Brewer, J.D. (2013) *The Public Value of the Social Sciences: An Interpretive Essay.* London: Bloomsbury Academic.

British Academy (2017) *The Right Skills: Celebrating Skills in the Arts, Humanities and Social Sciences.* Available at: www.thebritishacademy.ac.uk/publications/flagship-skills-right-skills-arts-humanities-social-sciences/ (Accessed:11/07/22).

British Academy (2020) *Qualified for the Future: Quantifying Demand for Arts, Humanities and Social Science Skills.* Available at: www.thebritishacademy.ac.uk/publications/skills-qualified-future-quantifying-demand-arts-humanities-social-science/ (Accessed: 11/07/22).

British Academy Response (2021) *Office for Students Consultation on Regulating Quality and Standards in Higher Education: Response from the British Academy* January 2021 (Accessed: 14/07/22).

British Academy Press Release (2021a) 25/01/21 *Latest Office for Students' Proposals Threaten Humanities and Social Science Courses.* Available at: www.thebritishacademy.ac.uk/news/latest-office-for-students-proposals-threaten-humanities-and-social-sciences-courses/ (Accessed: 14/07/22).

British Academy Press Release (2021b) 06/05/21 *The British Academy Responds to OfS proposals to cut funding for Arts Courses* Available at: www.thebritishacademy.ac.uk/news/the-british-academy-responds-to-ofs-proposals-to-cut-funding-for-arts-courses/ (Accessed: 14/07/22).

Bulaitis, Zoe Hope (2020) *Value and the Humanities: The Neoliberal University and our Victorian Inheritance.* London: Palgrave Macmillan.

Canterbury Christ Church University. *Future 360: The Canterbury Christ Church University Framework for Developing Enterprising Professional Graduates (2019–2024).* www.canterbury.ac.uk/career-development/docs/112-CH-19-CE-framework.pdf.

Cole, Doug and Tibby, Maureen (2013) *Defining and Developing Your Approach to Employability: A Framework for Higher Education Institutions.* York: HEA.

Dacre Pool, Lorraine and Sewell, Peter John (2007) The Key to Employability: Developing a Practical Model for Graduate Employability. *Education & Training*, 49 (4), pp. 277–289.

Daubney, Kate 'Teaching employability is not my job!': redefining embedded employability from within the Higher Education Curriculum'. *Higher Education, Skills and Work-based Learning* 12, 1 (2021) pp. 92–106.

Department for Education (2020) *The Impact of Undergraduate Degrees on Lifetime Earnings* Institute for Fiscal Studies Report by Jack Britton, Lorraine Dearden Ben Waltmann and Laura van der Erve Available at: https://ifs.org.uk/publications/impact-undergraduate-degrees-lifetime-earnings (Accessed: 12/01/22).

Department for Education (2022) *Higher Education Policy Statement and Reform Consultation.* Available at: www.gov.uk/government/consultations/higher-education-policy-statement-and-reform (accessed:01/03/2022).

Donelan, Michelle 'English Universities will soon carry a standard rating to stop poorer quality courses ripping off students'. *The Independent Newspaper* 20 January 2022 Available at: https://inews.co.uk/opinion/uk-universities-standards-rating-students-ripped-off-1413896 (Accessed:06/07/22).

Fazackerley, Anna (2021) 'Novelists issue plea to save English degrees as demand slumps' *The Guardian*, 19 June 2021 Available at: https://englishassociation.ac.uk/ea-media-guardian-190621/ (Accessed: 12/01/22).

Hazell, William (2021) 'Students could be prevented from taking university courses deemed 'low value' by the government'. *The Independent Newspaper,* July 22, 2021 Available at: https://inews.co.uk/news/education/students-places-university-courses-plans-cap-numbers-low-value-degrees-1116908 (Accessed: 12/01/22).

Heffer, Simon (2022) 'How Britain Abandoned its Classical Education'. *The Telegraph*, 6 August 2022 Available at: www.telegraph.co.uk/news/2022/08/06/how-britain-abandoned-classical-education/ (Accessed:14/09/22).

Helyer, R. and Lee, D. (2014), 'The role of work experience in the future employability of higher education graduates', *Higher Education Quarterly*, Vol. 68 No. 3, pp. 348–372.

Hubble, Sue and Paul Bolton (2018) *Higher Education Tuition Fees in England.* Briefing Paper 8151 Available at: https://researchbriefings.files.parliament.uk/documents/CBP-8151/CBP-8151.pdf (Accessed: 15/01/22).

Knight, Peter and Yorke Mantz (2004) *Learning, Curriculum and Employability in Higher Education.* London: Routledge.

McGettigan, Andrew (2015) 'The Treasury View of HE: Variable Human Capital Investment', 2015. Political Economic Research Centre Report. Available at: www.perc.org.uk/project_posts/perc-paper-the-treasury-view-of-higher-education-by-andrew-mcgettigan/ (Accessed 22/02/23).

McVitty, Debbie (2019) 'The Augar Review: The Essential Overview for HE' WonkHE. Available at: https://wonkhe.com/blogs/the-augar-review-the-essential-overview-for-he/ Accessed 22/02/23.

Moran, Joe (2022) 'Delivering the Undeliverable: Teaching English in a University Today'. *English: Journal of the English Association* 71 (273) 2022, pp. 1–22.

Morris, David (2017) ' A Beginners Guide to Longitudinal Education Outcomes (LEO) Data' WonkHE. Available at: https://wonkhe.com/blogs/a-beginners-guide-to-longitudinal-education-outcomes-leo-data/ Accessed 22/02/23.

Readings, Bill (1994) *The University in Ruins* London: Harvard University Press.

Reidly, Tess (2021) Hetan Shah, speaking in 'Arts Graduates are Flexible': Why Humanities Degrees are Making a Comeback' *The Guardian* February 16, 2021. Available at: www.theguardian.com/education/2021/feb/16/why-humanities-degrees-are-making-a-comeback (Accessed 17/07/22).

Shah, Hetan (2021) 'World View: Covid 19 Recovery: Science isn't enough to save us' *Nature* 591 (7851), p. 503.

Small, Helen (2013) *The Value of the Humanities* Oxford: Oxford University Press.

Tibby, Maureen and Norton, Stuart (2020) 'Essential frameworks for enhancing student success: embedding employability', Advance HE. Available at: https://s3.eu-west-2.amazonaws.com/assets.creode.advancehe-document-manager/documents/advance-he/EFSS%20Guide-Embedding_Employability_1589557033.pdf (Accessed 02/02/23).

Willets, David (2019) 'Graduate Earnings Rarely Afford Good Policy Making' *Times Higher Education*. 11 April 2019 Available at: www.timeshighereducation.com/features/graduate-earnings-rarely-afford-good-policymaking (Accessed 22/02/23).

Williamson, Gavin (2021) 'Education Secretary speaks at Launch of Digital Learning Review' Available at: www.gov.uk/government/speeches/education-secretary-speaks-at-launch-of-digital-learning-review 25 February 2021 (Accessed 15/01/22).

5 A fragile education for a good world

'Kenosis' and self-giving in teaching for sustainability and change

Ivan P. Khovacs

Sustainability is teaching for change

The scientist and philosopher Karl Popper famously declared that 'all life is problem solving'. The point of this chapter is to show the kind of problem-solving that putting the good of the planet on the university curriculum is for lecturers. I argue that it is both an ethical and pedagogical problem because we inevitably begin by asking: how can we get our students to take responsibility and care? How do we, from within the scope of our academic specialisms, help students apply moral conviction to problems of the environment? Is it even right for us to make that part of our job? And I show why this is where the presumption of academic objectivity in our field gives way to the social and relational subject, which is to say, where a pedagogy and ethics of the environment takes a personal turn.

To put it differently, the kind of problem-solving I am pointing to takes place within a tensile, multivalent, and interconnected world, in which, to cite David Orr,

> [a]ll of us are joined in one fragile experiment[:] we are all co-members of one enterprise that stretches back through time immemorial, but forward no farther than our ability to learn that we are, as Aldo Leopold once put it, plain members and citizens of the biotic community.
>
> (Orr 2004, p. xiii)

If this is the case, our fragile world needs educators as driven by our affinity for the world's beauty as we are by reason and the will to do what is good for a planet in peril. What the good is in the face of the environmental crisis depends, of course, on the kinds of creatures whose good we are trying to serve. Between beauty and tragedy, the good of the planet and its creatures depends on our moral perceptions that as humans we share in the fragility of the planet. No academic believes that the world we have been handed is one we would like to see remain unchanged. Discussions with fellow contributors in preparation for this book leave me in no doubt that we share a sense of

DOI: 10.4324/9781003286516-7

acdemic purpose: for each, the aim of scholarship and research is a renewal of vision, and teaching is always teaching for change. Bringing sustainability into the curriculum, however, throws up a moral dilemma: can we teach, say, facts about the mitigation of climate change, without invoking an ethical basis for action and change? Science gives us objective findings, but we need some means for making meaning from those findings. The facts of science do not supply their own answer to a question like 'why should we care?'.[1] That is a moral concern, and so is the now-familiar question 'what are we willing to give up for the preservation of biodiversity and zero-carbon solutions'.

I am arguing, therefore, that teaching for sustainability and care of the environment in higher education requires us to act as people who care, and that this is both a matter of pedagogy and ethos. Consider for example the fact that the World Bank estimates that by 2050, "without concrete climate and development action," 216 million people in sub-Saharan Africa, East Asia and the Pacific, South Asia, North Africa, Latin America, and Eastern Europe and Central Asia could be uprooted from their homelands and made refugees as a consequence of the growing impact of climate change (Clement et al. 2021). This statistic on its own has no power to compel the kinds of responses we need to reverse the potential devastation it portends: neither empirical knowledge nor know-how safeguard as a matter of course the behaviour change we need. Aristotelian virtues of prudence, justice, temperance, and justice in pursuit of 'the good' might be a starting point for reconsidering why we should care; so might religious values or humanist principles sourcing character, which is to say, our innermost sense of who we are and of the difference we intend to make with our lives, in the ethical act.

In the end, however we approach these questions with our students, it takes intellections running between heart and mind to perceive a wounded earth with any sense of moral agency and conviction that things can be different for us, for the planet, and for future generations. The reason is simple: as humans, we act in defence of what we love, whether land, self, community, or kin. Scaffolding a pedagogy of the environment on notions of love in response to a woundable planet is no exception: we teach from a position of self-involvement in what we love. As scholars and researchers, working with integrity of heart and mind means that academic detachment is a stance we can ill afford, indeed, by *acting as people who care*, we self-reveal to students when we teach. The ethos I champion in this chapter is in fact one most of us take for granted but already know in practice: the soul-searching, self-giving art of teaching rejects the trope that 'knowledge is power' and says, no, knowledge is love! Making sustainability central to the curriculum, therefore, requires openness with the students we mentor and teach about what we love and the ultimate good for which we aim.

This academic ethos undoubtedly puts us in a vulnerable position: does this not speak to a kind of fragility, a self-disclosing vulnerability in teaching? And does it not speak to the costliness of love, the price we pay in pursuit of what we love?

The fragility of teaching

In my philosophy of teaching, we change and at the same time are changed by what we love and long to know more fully. Teaching knows nothing of a detached, impersonal, immaterial solipsism in our intellect's hunger for transformation and change. Academics love their disciplines, their vocations, their scholarly communities, their students, their arguments, their books. To gain new knowledge is to love the pursuit of what is just, wise, good, demonstrably coherent, beautiful, and true. This takes passion, dedication, and an unflinching regard for what is evidentially true. It also requires making some kind of meaning from the knowledge we gain, so that we reach outside ourselves, transcend what is individual and temporal about ourselves, and inscribe our cognitions and self-reflexive experience into other consciousness: we write. This is why, in the humanities, mastering the intricacies of a new theory, mapping its thought patterns and implications, makes two conditions equally true:

(a) we intuit new knowledge in what is objective, external to us, and empirically true; at the same time,
(b) we intuit knowledge in art, ideas, and beauty, in nature and in the makings of social cohesion, because we seek in some sense to know ourselves and to know that we are known, we long to see in the world external to us ourselves.

This means that our very notion of what academic study is ought to start with the chord Socrates sounded out for his students long ago: to know is to 'know thyself', meaning that the pursuit of what we can know changes *who* we are and not only *what* we know. Put simply, isolating an object for study with dispassionate, unbiased objectivity does not mean that we engage with an inert, etherised subject; it cannot deliver us from problems which, in the end, interrogate us and get to the core of who we are. There is a difference between the Victorian collector of butterflies pinned in rows to a wall and the dedicated expert who takes herself into natural habitats, sees butterflies in their own environment, breathes the air they breathe, and in some sense learns to know herself as a subject in their world. Similarly, to stand before the students we long to see transformed by the scholarship we love is to be in some sense woundable, exposed, self-giving, and to that extent, self-revealing and undefended in our openness to change: to teach is to risk becoming ever more transparent to students and to ourselves.

This chapter explores these ideas in a critical reflection on an experiment in bringing sustainability into a practical theology classroom. In what follows, I offer an illustration and critique of what an immersive pursuit of what we love might look like, on the premise that this is how we teach our students to be changed by that which is worth their attention and their love. The point of this chapter is, in fact, methodological: I seek to sketch by way of a case study a philosophy of self-giving in teaching for sustainability and change. The hypothesis I am testing is that getting students to draw on what they care about and love is a necessary condition for putting environmental action on the theology and philosophy agenda.

The module that forms my case study involved a curriculum redesign putting food at the centre of ethical action for the planet, a change I anchored on the university's stated aims for sustainability and 'greening' the curriculum. The critical analysis on this will, I hope, bring home the point I am making about moral value and self-knowledge, and why these constitute a particular kind of good students can expect from higher education. I will conclude with the dual conviction I have gained in the process: namely, that loving a world made increasingly fragile by the harm we cause the environment means turning our cognition of our sharing in that fragility into an act of self-giving for our students, and indeed that this *kenotic* act of head and heart is teaching for sustainability at its best.

'Greening' the curriculum

I have been teaching this module for just over ten years, relating questions of faith and spirituality to a hermeneutics of care and compassion sourced in a broad and inclusive body of Christian scholarship. The stated aims of this module have always been to enable students to explore their own values, whether secular, spiritual, or otherwise, following principles of practical ethics and the pursuit of self-knowledge outlined above. We do this by testing case studies against analytical readings within a broad Christian frame of symbols and sources of critical reflection. Theology and Religious Studies undergraduates (invariably representing a spectrum of beliefs, faith traditions, and none) combine their work – in critical methods of reading sacred texts and the historical development of religious philosophies and creeds – with a contemporary focus on religion and its entanglements in our day. A surprising number of students, about 20 per cent of any given year group, have gone on to become ordained ministers in the Church of England, though the course is not in fact designed around vocational aims. Another 40 per cent or so go on to teach Religious Education in the schools. Others pursue careers in the charity sector, overseas development, or in jobs at the intersections between social justice, lived religion, community advocacy, and the arts.

In a typical semester, this practical theology module requires students to devise action plans, applying a modified version of Kolb's reflective learning cycle, in response to pastoral case studies drawn from the practical theology literature. At its simplest, practical theology is where theory meets practice, its "research methodologies take account of the 'storied' and hermeneutical nature of human culture," (Graham 2017, p. 175) and so puts us in touch with the spiritual and ethical nature of human flourishing; this relies on bringing beliefs, values, and virtues into dialogue with sources of human motivation, meaning, and purpose. Remarkably, these final-year undergraduates, when presented with sound principles of care, tend to demonstrate maturity and ability to navigate (albeit in the experimental confines of the case studies) the emotional and ethical complexities of attending to personal conflict or to lives in distress, and generally to situations calling for deep resources of compassion and the skills to show care in practical ways. We typically use the parable of the Good Samaritan as an invitation for students to reflect critically and creatively on the centrality and limits of compassion, the role of emotions, the inequalities and power dynamics in offers of care. Learning to take a non-judgemental evaluative stance is key to this work, and so are the challenges of self-care in helping others wrestle with complex ethical and emotional conflicts.

The case studies we work with might involve extending practical help and spiritual support for persons in hospital, first-person accounts of refugees and asylum seekers, or meeting persons with profound learning disabilities and learning from the challenges with which they live. Core to the assessment strategy are reflective practice portfolios; these are often deeply moving accounts of students marked by personal transformation expressed in terms of students' changed outlook or a freshly unfolding understanding of what in the end is of value to them as growing, maturing persons learning to live lives of meaning in what they can do for others. Above all, the emphasis on experience, ethics, and action serves the pursuit of self-knowledge in this process. With this in mind, in preparing the year's syllabus, I determined that the learning outcomes were written broadly enough that I could overhaul the content with a practical theology of food and ethics of the environment focus.

I have developed elsewhere the principle that, as a Christian ethos of sustainability and care of the environment, 'green theology' is simply theology (Khovacs et al. 2022). In the contemporary scholarship, the language of 'stewardship' has been used to articulate an ethic derived from the biblical notion that God contemplates the world with love and in doing so enjoins humans to contemplate the world with awe, compassion, and in some sense love,[2] but in the end also to turn this into action and so to tread the earth with care, even as the first humans are said to have "heard the sound of the Lord God treading in the garden in the cool of the day" (Genesis 3.8a). In the biblical origin story, first humanity is tasked with tilling the ground and perpetuating God's goodness in creation. If origin stories are intended to convey universal meaning, Genesis tells us that human activity of sowing, tending, and gardening the earth are intended to mirror the creativity and care with which God fashions and contemplates the world. In recent decades, the term 'stewardship' has been used to characterise the human vocation to care for the planet, it is a commitment premised on the idea that 'and God saw, and it was good' is the loving prelude from which God raised everything to new life. For some, however, the term stewardship is compromised by an implied hierarchy putting humans above the rest of the natural world.[3] Our students generally agreed with critiques of the 'stewardship' model, and unanimously rejected uncritical "assumptions that some crucial quality radically separates humans from nonhuman animals and 'nature' generally" (Peterson 2001, p. 2). They agreed, on the other hand, with

Holmes Rolston III, one of the founders of environmental studies, for whom the Judeo-Christian creation narrative resonates with contemporary notions of human accountability for nature. For Rolston, in the biblical text, "*Homo sapiens* is the aristocratic species on earth" (Rolston 2020, p. 67), but in such a way that sets up some distinct ethical parameters in which to be human is to image back something of God's care for creation:

> *Humans* are cognate with the *humus*, made of dust (as *Genesis* teaches, long before any bioscience), yet unique and excellent in their aristocratic capacity to view the world they inhabit. They rise up from the earth and look over their world (Greek: *anthropos*, to rise up, look up). Persons have their excellencies, and one way they excel is in this capacity for overview. [In this sense,] they are made "in the image of God."
>
> (Ibid., p. 58, original emphasis)

The context to this argument is the fact that for much of history humans have corrupted the biblical vision Rolston points to with opposing ideas of nature's subjugation and domination used to justify our abuse of the natural world. David Orr, among others, has joined in the critique that Christian beliefs have fuelled hierarchical, egocentric, anthropocentric, and specifically androcentric (male-centred) and destructive views of nature; he notes: "the writers of Genesis commanded us to be fruitful, multiply, and to have dominion over the earth and its creatures. We have done as instructed" (Orr 1992, p. 11). Our students, however, equipped with the tools of hermeneutics and literary analysis were quick to point out that sometimes context is everything. They pored over the creation texts in Genesis, paying attention specially to literary and contextual Hebraic meanings of terms like 'subdue' and 'dominion' which, read on their own without context, carry connotations of violent conquest.[4] This is why Bible translations contextually doing justice to the letter and agrarian spirit of Genesis[5] use terms like 'govern' or 'bring under control' to account for the ethical implications of human stewardship over the natural world. Consequently, students concluded that having an overarching perspective over the natural world makes humans especially accountable for the welfare of the planet, it does not make humans nature's rulers whose endless wants, earth's 'resources' are there to fulfil.

Students were in fact uncompromising in their reasoning that, in the six days of creation in Genesis, land, rivers, and sky, animals, plants, and creatures of the sea all precede humanity's arrival on the scene, and that this meant that the non-human world has its own integrity, beauty, and goodness, and is uniquely cherished by God. They reasoned that the leitmotif 'God saw, and it was good' which accompanies each of the days of creation imbues on the late-arriving humans a sense of belonging to the natural world. Clearly, though, what we imagine is included in 'nature' and the wider web of 'creation', and what meaning we believe we can make from this, matters. In the Genesis creation myth, humans are clay with a fate strangely seeded in forbidden fruit. This says something about the story's original exponents, namely, people living on an ancient Mesopotamian plain and perennially defending crops against floods, droughts, plagues, and other ravages of nature: for these humans, labouring on the land was their ethic and story, eking out their livelihood from the ground and learning to live in a fruitful, regenerative, and meaningful relationship with the natural world was both existential ethos and humanity's essential task. Surging forward in the world with some sense of purpose beyond the day-to-day, therefore, meant knowing that humans belong to the whole of the plant and animal kingdoms, and by the same principle also belong to land, rocks, rivers, mountains, and so on. An ecological reading of Genesis, there-fore, would tell us that to be human is to live with a sense of moral obligation for the custodianship and care of everything God has made. With this philo-sophical and ethical tool kit at their disposal, students launched into our study with the conviction that

(a) to be human is to be the storytelling, meaning-making creature which countless origin stories, and Genesis among them, attest to;
(b) to derive meaning from and tell a story binding humans to their origins in nature makes humanity responsible for behaving as guardians of the nat-ural world, not its masters.

This is the sense in which, in our module, a 'fragile education' meant helping students to know what they love and to love what they could account-ably pursue in rendering a future that is good, sustainable, and worth striving for with hope. How did this then impinge on a food ethic and care of the environment?

Extending the concept of care: peeling a potato

"When was the last time you peeled a potato?" This was my opening gambit in introducing the module's revised focus on food ethics and sustainability, extending the concept of care to an ethic of the environment. The question was part of a survey designed to show students how little we understand the food we eat, where it comes from, and the impacts our food choices have on the natural world. As far as many are concerned, food comes packaged on a store shelf. The chain of human labour and the preparation processes connecting the meal on our dinner plate to the soils and waterways which become our sources of food can be largely a mystery to those of us who no longer live near farms and the traditional sources of food production. As the poet, theologian, and farmer Wendell Berry observed,

> in the advertisements of the food industry, ... food wears as much makeup as the actors. If one gained one's whole knowledge of food from these advertisements (as some presumably do), one would not know that the various edibles were ever living creatures, or that they all come from the

soil, or that they were produced by work. ... The products of nature and agriculture have been made, to all appearances, the products of industry.

(Berry 1990, pp. 147–148)

Asking when students last peeled a potato was only the start of a wider conversation about how contemporary eating habits have severed our connection to the land and to the natural world generally. We were then free to explore whether a 'practical' theology could rekindle practices of grateful and ethical eating. Our talk about food was, of course, a useful laboratory model from which we could raise a critical question in response to the ecological crisis: could we relate our love of food to spiritual values and to an ethics of the natural world we want to save? These changes in perspectives and food choices for the good of the environment helped us to consider more difficult questions: why should we bother? Why and how should each of us show that we care? Or, from the point of view of our classroom, could a multi-ethnic, middle-to-modest income group of students representing an essentially urban class, recognise their agency in the wellbeing of landscapes, water sources, and soils, and of the fragile interdependent biological structures sustaining life on the planet?

Our critical reading included the pioneer environmentalist, Aldo Leopold, who rejected the anthropocentric view of nature, but rejected also a purely biocentric argument that puts all life, animal, plant, and human, morally on an equal plane. Biocentrism's appeal is twofold: one, it appears to be free from the irrational, intuitive, and imagistic reflex of religion; two, in consequence, any sense of moral meaning derived from nature is conditioned, not just by human ends and lifecycles, but by the mechanisms of interdependence between all species, and so by the needs of species complexity and diversity. Despite the appearance of a rational, science-driven means to ecological ends, however, Leopold's critique of biocentrism cut to the quick: by giving humans no greater or lesser importance than other animals or plants, biocentrism denies human obligation to other living things, namely, by setting up a Darwinian competition for diminishing resources. Leopold advocated instead a "land ethic," a "biotic," cyclical view in which land is not dead, inorganic dirt, but substrate and lifeline for successive layers of living things, each layer becoming kenotic (self-giving) life giving off energy to sustain the next (see Figure 5.1):

Land, then, is not merely soil; it is a fountain of energy flowing through a circuit of soils, plants, and animals. Food chains are the living channels which conduct energy upward; death and decay return it to the soil. The circuit is not closed; some energy is dissipated in decay, some is added by absorption from the air, some is stored in soils, peats, and long-lived forests; but it is a sustained circuit, like a slowly augmented revolving fund of life. There is always a net loss by downhill wash, but this is normally small and offset by the decay of rocks. It is deposited in the ocean and, in the course of geological time, raised to form new lands and new [biotic] pyramids.

(Leopold 1987, p. 216)

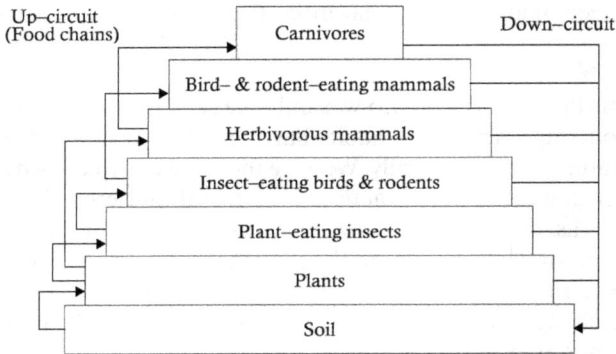

Figure 5.1 Biotic pyramid, showing plant and animal community as an energy circuit.

Source: Adapted by author from published notes by Leopold (1939).

Leopold's view that land is a living, cellular organism from which plant and animal structures draw their energy in an upward/downward flow speaks to the importance of the health of soils for sustaining the fruitfulness of plants, insect varieties, and the proliferation of bird life and other natural pollinators, a principle which household gardeners know well. The health of lands is by the same token dependent on complexity in "the characteristic numbers, as well as the characteristic kinds and functions, of the component species" in the supporting chain structure (Ibid.). Failure in one part of the structure threatens the collapse of the whole. This built-in interdependence making possible the diversity and complexity of life, should make us think twice about dirt, sod, soil, loam, and so on, for whatever else it is, and whether we speak of it as an "energy unit" (Ibid.) or the engine room of life itself, dead matter, it is not. This ethical affirmation was of course the point of pairing creation stories with stories about the food we eat.

Pedagogically, to bring the point home, the pinnacle I needed to climb with our students was to rethink attitudes about the natural world with a challenging proposition along these lines: what difference would it make to believe that all humanity 'belongs' to one common soil that sustains all organic life and inorganic matter alike? I assigned student presentations exploring the idea that food acts as a conduit of relationship between humans and land, animals, plants, water, air, and so forth. The students gave 30-minute presentations on food, investigating food histories of their choice, outlining their respective cultivation practices, and fleshing out the cultural significance surrounding the kinds of foods we take for granted: coffee, chicken, chocolate, and cheese, were among the students' favourites. Taking their findings on food to their logical ends, the students were emboldened to take up the fragmentary call from Genesis and to re-narrate their own fragmentary ontology of belonging as an ethic of commonality with the ground beneath our feet. Students articulated with their own eloquence a kind of philosophical anthropology in which every act of eating is a retrieval and re-enactment of our shared 'creatureliness' with other creaturely life; ultimately, they made the re-enchantment of eating a life-ethic in which taking a meal is both an act of communion with the earth and an opening to our apprehensions of the transcendent 'other' in our fellow human.

One student gave a presentation exploring 'The Human Side of Rice'. The presentation included a brief cultural history of rice, a discussion about its common varieties originating in Asia some 30,000 years ago, its propagation throughout the planet, rice in global economic activity, rice as a staple for much of the world, and concluded with questions about of rice cultivation and global warming. The student gave examples of colourful and variously spiced rice preparations from around the world. Fellow students were surprised to learn that rice is a semi-aquatic plant that requires all-season irrigation conditions, hence why it is grown in clay and silted loams heavy on water retention. Students learned that this type of farming is often done by hand and, depending on geography and culture, it might be work done by women. They also learned that there are more than 120,000 varieties of rice in the world, typically classed by place of origin, kernel size and shape, starch content, and flavour. Students were deeply moved by images of women, often also children, in either broad brimmed hats or other practical head covering at work knee-deep in flooded rice fields. They expressed in the ensuing discussion a sense that something had changed, that the idea of eating rice was imbued, as it were, with a layer of humanity, a sense that human hands and a long history of agricultural practices coming from deep knowledge of soils and seasons are behind every grain of rice they ate from a takeaway tray.

Another student spoke about fruit varieties from her family home in the Caribbean, bringing into question what we mean by 'foods we take for granted', given that the fresh fruits she is used to on holiday visits to family are not found on store shelves in the UK. Another student gave a presentation on a cooking technique from the Hadhramaut region in Yemen, in which meat and rice flavoured with a blend of spices is cooked on hot rocks in an underground pit. The student then drew the class into a discussion about this food tradition in the context of the present famine in Yemen: students concluded that the starvation of bodies inevitably means the starvation of land, and with that, of culture, of family and community celebrations around traditional 'mandi' cooking.

As a point of contrast, a student supporting herself with work at Starbucks, delivered a presentation on the corporate 'sameness' in the company's coffee shops she had visited in Eastern Europe, Asia, and the United States. The student concluded that Starbucks presents coffee as a global drink, but that this means, effectively, that Starbucks markets two products: coffee, of course, but also *the idea of coffee*, the idea of bringing people together in the common enjoyment of taking coffee. As the student saw it, the idea is the more important product of the two. The shared enjoyment of the Starbucks product is also given a social context in shops which, rather like airport lounges, operate as neutral social spaces: Starbucks restaurants the world over are styled to look essentially the same, without overt identification with the host culture, and designed to function interchangeably as both work and leisure space. Students detected in this spatial and aesthetic arrangement the engineering of a cultural 'amnesia', an erasure of the roots that as humans we share in the tropical African terrains where the *coffea* plant originates. Students concluded that the starry-eyed siren in the company's ubiquitous green logo is an invitation to regard the human labour making it possible to enjoy a bean-to-cup experience strictly from a safe distance, as is evident, for example, in images posted around the shop of healthy, smiling coffee growers who make no moral demands on the Starbucks consumer, whether about fair labour and trade practices, or about the costs to soils and to crops cleared to make room for the highly profitable *robusta* and *arabica* coffee plantations. Students were also left uncomfortable with questions about the cost in biodiversity to coffee varieties and, culturally, about the displacement of other forms of drinking coffee that comes with the one-size-fits-all Starbucks experience.

In sum, exploring the biblical creation narrative (alongside others such as the Enuma Elish and Haida creation myths)[6] had given the class a key critical standpoint: creation stories tell us that humans are not God, we did not come from ourselves, and we belong to the world outside ourselves. Framing our discussion of food around creation myths, therefore, served a dual purpose: first, it enabled students to think of humanity's common roots in soil, geography, and landscapes by considering various foods and their rooting in the cultures and histories of people in places across the globe. This enabled the class to reflect on our contemporary amnesia erasing human connection with land in the light of an ancient and sobering judgement calling us back to the earth: "remember that you are dust and to dust you shall return" (Genesis 3.19). Second, as a study in ethics, the exercise was a radical critique of anthropocentric hierarchies, whether secular or theological, in which humans see themselves as both centre and endpoint of all earthly purpose. The point was to enable students to be challenged by a sense of mutuality and interdependence on what nature is, and to grow from this a sense grateful living for the benefit humans have learned to reap from a balanced relationship with the non-human and the material world. The presentations on foods and the cultivated environments in which they grow gave the students the rationale they needed to affirm as a matter of human and in some sense of spiritual principle our common belonging to nothing less than dirt.

Fragility as an ethos for sustainability in higher education

In leading this module, one concern for me was whether confronting students with the catastrophic collapse of ecosystems – scarcity of resources, humanity's apparent inability to come to a mind on climate change, and little evidence that the nations responsible for the highest levels of environmental damage can commit to costly policies and change – would, in fact, stimulate students to self-critical discussion leading to possible solutions. My problem, essentially, was this: faced with environmental doomsday scenarios, would students lose their sense of agency to pessimism, and so, give in to feelings of futility, shirk responsibility, and simply give up? Or, on the other hand, could students be motivated, not by disaster planning, but by building on what they love, and fleshing out notions of the difference they could make in the world? On this point, the Canadian farmer and theologian Norman Wirzba offers a devastating critique of zero-sum interpretations of the Darwinian evolutionary thesis: Wirzba warns against drawing conclusions from the biological, evolutionary model about how we ought to behave towards one another and other creatures given the forces of competition, scarcity of resources, and the need for lifeforms to assert superiority over one another. For Wirzba, Darwinian evolutionary science can provide the underpinnings of a moral theory of environmental responsibility, but we must begin by asking a different set of questions, we cannot let our ethical frame be reduced to the level of a fight for survival:

> What does [evolutionary theory] leave out of view and out of consideration, and what might it prevent lookers from seeing? Why put the focus on competition among individuals rather than cooperation among groups? Why assume scarcity in a world that might also be characterized by great abundance? Why believe that the drive to live is a drive primarily to "survive" rather than to "thrive" or "delight"?
>
> (Wirzba 2022, p. 92)

The point I needed to emphasise to the students was that we long to thrive, not merely survive. This was an insight we explored with Wirzba, for whom "[t]he future of a healthy and vibrant world depends on people seeing, hearing, smelling, touching, and tasting fields, forests, waterways, and fellow creatures *as gracious gifts* and not merely as units of production or consumption" (2022, p. 23). Wirzba sharpened our focus on the ethics of sustainability given our consumption, sheer need, desire for, and enjoyment of, food. Humans, in other words, do not eat merely to feed an animal hunger but eat and take delight in the food we eat. We share food as an extension of our emotional history, not just to keep existing, but as an affirmation of life itself. If this sounds rather ambitious, however, perhaps we can look at it in a different way. Of all the work these final-year students had done in their theology, religious studies, and ethics degrees, writing about their love of food and reflecting on what it means to become grateful eaters may just have been the most radical act of resistance against a culture of industrialised food – a culture that threatens to cut us off from our conscientious, generous, and gracious participation in *the wider web of life which gives itself for other life*, in the biotic complex in which, one day, we too will become food for the earth.

And so I come to a concluding defence of the term 'fragility' as an ethos for sustainability in higher education.

There is an ethical concept that always sounds far-flung and abstract on first encounter, the term *kenosis*, a Greek word meaning 'self-emptying'. The word in its dictionary form sounds intimidating —What am I being emptied of, and why?—until we realise that the term simply means 'letting go'. As a piece of religious reflection, *kenosis* can denote a spiritual ideal, choosing a path of renunciation in monastic life, for example. In Christianity, *kenosis* can also signify a posture of humility, not as an external imposition (by forms of oppression) but as an ethical disposition in which I choose a path away from privilege and power. If I choose self-abnegation and opt instead to identify with poverty, with those on the margins, with vulnerability in places of social exclusion, that is a practice of kenosis. The importance of humility for the Apostle Paul led him to go so far as to say that 'to die is gain' (Philippians 1.21). Kenosis means loss, but a loss that can only be measured against a greater moral or spiritual good to be gained as a result.

I would like to suggest that academics and educators know something of this path of renunciation, or simply self-giving, even without the term. At some point in the life-long formation of the teacher, we are confronted with the fact that the pursuit of what we love requires letting go of the idea that 'knowledge is power'. At that point, knowledge acquires for us a quality that far surpasses in value the sheer volume of information we can pour into people's heads. In its place, we begin to favour an educational ethos in which knowledge is not just 'subject expertise', but a particular kind of love nurtured on that which is most true about ourselves. This is where we are confronted with the fragility implied in the question: what do we love? Every geographer, biologist, Shakespearean, philosopher, or anthropologist faced with students, whether incomparably eager or seemingly indifferent in their endeavour, has met the challenge of the classroom by rehearsing to themselves questions of deep human value: what, in the end, do we care about and think is worth investing ourselves in with empathy, foresight, and vision? To what do we give our curiosities, spiritual attention, moral dispositions, appetites, and intellectual drives? And why? *What do we love?*

In the end, I am defending the costliness of love, love of the planet, of our expertise in certain subject areas and, finally, of students and of the values that drive us in higher education. I have characterised the outworkings of this ethos as 'fragility', a kind of self-giving, denoting that which is necessary to let go of in order to more ardently pursue what we love. I use fragility in a positive and radically empowering sense, as a term advocating an educational ethos trained on the flourishing of our fellow other. A 'fragile education', I argue, is a concrete expression of St John the Divine's ethic which prizes love above all, an ethos which puts the perception of particular people and situations above abstract principles and rules: act out of love, says John, for "God is love" (1 John 4:16).[7]

And since it is about love, as a morally compelling and empowering ethos, a fragile education makes sustainability a proposition irreducibly about persons and not primarily about abstract ideas, let alone instrumentally about what we acquire in the process. Even to speak of 'the environment', then, is to prioritise biotic alignments in which we exist as thinking and feeling human persons in response to, and with responsibility for, other living creatures and things. To quote, once again, environmental educator David Orr, "were we to confront our creaturehood squarely, how would we propose to educate? The answer, I think, is implied in the root of the word *education, educe,* which means 'to draw out'. What needs to be drawn out is our affinity for life" (Orr 2004, p. 213).

At its simplest, then, I am working out a grammar of compassion and care making teaching for sustainability not a transactional activity with abstract knowledge as its currency, but a relational, ethical, and regenerative good.

From this standpoint, the students in our practical theology of food and sustainability study could envision a world in which their eating habits become inscribed into patterns of mutual flourishing with the non-human and with the very *humus* from which all things originate. 'Remember you are but dust'. And they could imagine a world in which 'dominion' does not have to mean destruction, and nature is not an enemy to be subjugated and conquered. They could imagine a world in which human evolutionary superiority does not result in species extinction, or in the industrial-scale effluence causing climate change.

As a practical theology, giving students agency to reimagine their place in the natural world empowered them to conclude that acting in defence of nature is a fraught but profoundly human disposition.

If we see ourselves as earth's thinking, self-conscious organism, if we are the moral coinherence of heart and mind in a universe that believes in itself, then we are always faced with moral choices. And so we can choose to act in defence of the environment, not as a salvage operation of last resort, but because we are built to act with hope and on behalf of what we love. And so, I end where I began, by asking: how can we help our students know what they love and name that which will lead them, time and again, to choose life?

Conclusion

I have argued in this chapter that the change we want to see in our students is what makes us irreducibly the educators that we are. This inevitably makes us vulnerable and ethically transparent about the good we long to see mirrored back to us in our students. The inchoate, fragile, kenotic, self-revealing 'I' that I am depicting may be the most teacherly gift we can give to our students. We are the mirror in which students can see themselves becoming self-disclosing about what they love, and through which they can learn to love and have faith in what they do not yet know; we are the lens in which they can see gradually pulling into sharper focus accounts of ecological healing and hope for a world in which they can believe.[8]

Notes

1 David Hume's classic argument that you cannot derive an 'ought' from an 'is' – that is, an ethical moral imperative from mere facts – applies here. For Hume, any change compelling moral obligation ('what I ought to do') from a mere description of how

things are depends entirely on the introduction of a moral requirement or precept (whether God, tradition, law, nature, self-preservation, universal human rights, etc.). This may well happen by tacit ascent rather than explicit assertion, but that is all the more reason why Hume warns that the change to 'ought' from 'is', however "imperceptible," says Hume, is a matter of ultimate "consequence." (Hume 1978, p. 469).

2 On the question of whether we can 'love' the planet and what this does for an ecological ethos from a religious perspective, see Fiddes (2023), especially 'Outlines of an Eco-Love-Theology', pp. 19–224.

3 Steven Bouma-Prediger (Hope College, Michigan) argues that "ecological literacy and good earthkeeping" (p. 90), as well a sensitivity for the holy, shows that the stewardship model unhelpfully "underwrites a dualism" of human culture and purpose over nature, and therefore "sanctions only a managerial role for humans" (p. 85); cf. his chapter in Warners & Heun (2019), pp. 81–91. The term 'earthkeeping' is Loren Wilkinson's: cf. Wilkinson (1991). Ruth Valerio (community activist, academic, and Global Advocacy director for the sustainable development charity Tearfund) objects to the commodification of the natural world, and our treatment of the wider web of life as property to be managed economically; she calls for an ethics that eradicates the instrumental and patriarchal associations with the 'stewardship' model (cf. Valerio 2020).

4 The pernicious connotations of dominion theology is something that present scholarship has on the whole been able to disown in view of countervailing biblical notions of God's redemptive love of the world God creates. Cf. Middleton, J.R. (2005), p. 51.

5 On the agrarian spirit in Genesis, see Norman Wirzba, for whom the creation poem, as he calls it, says that "a human life is not separable (from the ground) and self-contained but is instead an extension of, and in constant need of, the ground out of which it arises and by which it is daily nurtured" (Wirzba 2022, p. 51).

6 Compare the eco-theological ethos in the biblical narrative with one from the indigenous origin myth of the Salishan people of the Pacific Northwest in the United States and the Columbia Plateau of Canada. In that story, Coyote is sent to earth to teach people the arts of living well with the land before returning to his home with the Old-One, for "If matters are not improved on earth soon, there will be no people" (Staiger Gooding 1992, p. 258).

7 On love as the principle of philosophical and ethical insight, see the work of Martha Nussbaum, especially Nussbaum (1998), pp. ix, 256–257; also Nussbaum (1986), pp. 12–17.

8 I wish to dedicate this essay to Professor Loren Wilkinson (Emeritus, Regent College, Vancouver), who has helped to shape much of my ethical thinking, scholarship, and Christian imagination. Cf. his forthcoming Wilkinson, L. (2023). *Circles and the Cross: Cosmos, Consciousness, Christ and the Human Place in Creation*. Eugene, OR: Cascade Books.

References

Berry, W. (1990). 'The Pleasures of Eating'. *What Are People For? Essays*. New York: North Point Press.

Bouma-Prediger, S. (2019). 'From Stewardship to Earthkeeping: Why We Should Move Beyond Stewardship'. *Beyond Stewardship: New Approaches to Creation Care*. Warners, P. and Heun, M., eds. Grand Rapids, MI: Calvin College Press.

Clement, V. et al. (2021). 'Groundswell Part II: Acting on Internal Climate Migration'. Washington, DC: International Bank for Reconstruction and Development/The World Bank; sourced 6 February 2023 on http://hdl.handle.net/10986/36248.

Fiddes, P., ed. (2023). *Loving the Planet: Interfaith Essays on Ecology, Love and Theology*. Oxford: Firedint Publications.

Graham, E. (2017). 'The State of the Art: Practical Theology Yesterday, Today and Tomorrow: New Directions in Practical Theology'. *Theology*, vol. 120, Issue 3, May–June 2017; pp. 172–180.

Hume, D. (1978). *A Treatise of Human Nature*. Oxford: Clarendon Press.

Khovacs, I. et al. (2022). 'Land as God's Common Good: Reframing Territoriality with a Theology of Land as Shared "Creation."' *Doing Climate Justice* (Theological Explorations: Religion and Transformation in Contemporary European Society), vol. 21. Paderborn, Germany: Brill Schöningh.

Leopold, A. (1939). 'A Biotic View of Land'. *The Council Ring*, vol I, Nov. 13, 1939. Washington, DC: Department of Interior, National Park Service.

Leopold, A. (1987). *A Sand County Almanac and Sketches Here and There*. Oxford: Oxford University Press.

Middleton, J. R. (2005). *The Liberating Image: The Imago Dei in Genesis 1*. Grand Rapids, MI: Brazos.

Nussbaum, M. (1986). *The Fragility of Goodness: Luck and Ethics in Greek Tragedy*. Cambridge: Cambridge University Press.

Nussbaum, M. (1998). *Love's Knowledge: Essays on Philosophy and Literature*. Oxford: Oxford University Press.

Orr, D. (1992). *Ecological Literacy: Education and the Transition to a Postmodern World*. Albany: State University of New York Press.

Orr, D. (2004). *Earth in Mind: On Education, Environment, and the Human Prospect*. London: Island Press.

Peterson, A. (2001). *Being Human: Ethics, Environment, and Our Place in the World*. London: University of California Press.

Rolston III, H. (2020). *A New Environmental Ethics: The Next Millennium for Life on Earth* (2nd edition). Abingdon, Oxon: Routledge.

Staiger Gooding, S. (1992). 'Interior Salishan Creation Stories: Historical Ethics in the Making'. *The Journal of Religious Ethics*. Vol. 20, No. 2 (Fall, 1992); pp. 353–387.

Valerio, R. (2020) *Saying Yes to Life*: London: SPCK.

Warners, P. and Kuperus Heun, M. (2019). *Beyond Stewardship: New Approaches to Creation Care*. Grand Rapids, MI: Calvin College Press.

Wilkinson, L., ed. (1991). *Earthkeeping: Stewardship of Creation*. Grand Rapids, MI: William B Eerdmans.

Wirzba, N. (2022). *Agrarian Spirit: Cultivating Faith, Community, and the Land*. Notre Dame, IN: University of Notre Dame Press.

Begins with a sigh

Victoria Field

Sustainability is a word containing other words just as Canterbury contains the dramas of its invasions and upheavals. Sustainability is a flowering, rooting, slithering word that can also shake out its wings and fly. In its current usage, it's only the same age as me, born in the early sixties. But its six-syllabled architecture has the structures and significances of old French and Latin, both languages at one time the dominant tongue spoken in our city of Canterbury. Sustainability begins with a sibilant sigh and ends jauntily with its roots in *habilis*, something handy and manageable.

Acknowledgements: parts 'Begins with a sigh' are taken from Field, V. (2020) *Not Utopia … But Maybe*, Canterbury: Canterbury Christ Church University.

DOI: 10.4324/9781003286516-8

Part 2

Pedagogies of (re)connection

Pedagogies of (re)connection explores the potential of holistic approaches to teaching and learning that can support the development of complex relationships between the human and more-than-human worlds. The following chapters focus on the significance of embodied experiences with otherness, highlighting the importance of considering how the individual is implicated in such interactions and what their responsibilities and learning opportunities for a better world might be.

DOI: 10.4324/9781003286516-9

6   A sense of beauty in belonging to the whole

Tansy Watts

Introduction

My engagement with holistic philosophy initiated in Early Childhood Education (ECE), has been developed through post-graduate study, and I am currently exploring its application to teaching in higher education (HE). The holistic paradigm orients education in support of the growth, renewal, and development of the individual in relation to the whole (Mahmoudi et al., 2012). This educational vision starts at the point of connection between all life and sits in opposition to that focused on individual development. Indeed, the holistic paradigm engages with the need to move beyond a focus on individual development and engage with the equally significant dimension of learning how we "belong" to a whole ecosystem (Davis & Elliot, 2014). My journey to this perspective has been deeply personal, and in this chapter I share insights from this to illustrate and explain the "whole-part" relationship. My interest is focused on ways of knowing and their impacts on relations, from early to higher education and beyond.

Higher education has been shaped by a Western intellectual legacy and its valuing of "reductionism, objectivism, materialism, dualism, and determinism underlain by a mechanistic metaphor" (Sterling, 2021, p. 5). However, it is these values that are associated with the creation of "an unsustainable and degenerative relationship with the ecosphere," and their perpetuation through HE as part of the problem. The naming of the current era as the Anthropocene can serve as the means to disrupt a persistent "humanist paradigm" (Malone, 2018, p. 4) and support the potential for "risking finding new ways of relating to the world" (Taylor & Pacini-Ketchabaw, 2015, p. 510). Shifting the paradigm underpinning higher education is recognised as fundamental to its repurposing towards human and environmental flourishing. Although the scale of change required is great, Meadows (1999) suggests that such shifts can occur individually within "a millisecond. All it takes is click in the mind, a falling of scales from eyes, a new way of seeing." (Meadows, 1999, p. 18). It is on this basis that I now share insights from my professional and personal experience and explore the relevance of early educational relational pedagogies for higher education.

DOI: 10.4324/9781003286516-10

78   *Tansy Watts*

A vocabulary of holistic relations

I start by introducing Jean Gebser's evolution of human consciousness theory, which I have found useful in understanding a holistic perspective. It is based on a spiritual conception of life that is specifically non-religious, but which recognises an atemporal, immaterial source of all life referred to as "Origin" (Gebser, 1949). The vocabulary of holistic relations is presented as five broad, but distinct descriptors of relational qualities: *archaic, magic, mythic, mental,* and *integral consciousness*. These descriptors offer the means to give equal value to *experiential, affective, creative,* and *abstract* relations. Holistic awareness lies in the potential to integrate these different forms of consciousness. The following diagram (Figure 6.1) offers a visual aid for considering the rich, continuous, connective nature of such multi-dimensional relations.

This theory aligns with other contemporary perspectives that draw on a sense of life's sacred nature when contemplated in whole or collective terms, including post-human and new materialist perspectives. Braidotti, for example, draws on the ancient Greek concept of "zoe" in considering all life materials as imbued with vitality, and the universe as one infinite and indivisible "dynamic, self-organizing structure" (Braidotti, 2013, p. 60). This concept has been used to promote the potential for a "zoe-centric" worldview as an ethical model that values this life force centrally and can thereby displace the human. Engaging

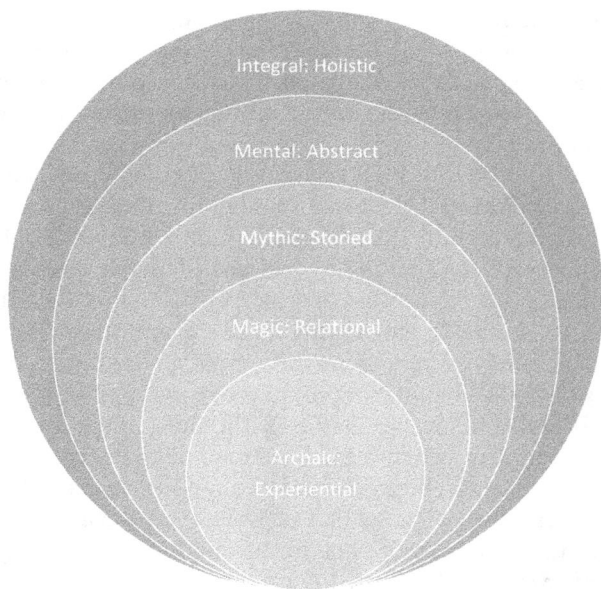

Figure 6.1 A vocabulary of holistic relations.

Source: Gebser, 1949.

with such perspectives can offer a means to step beyond the influence of dominant cultures and scope for "new ways of thinking about our place in history, and what our potential for future evolution might be" (Yiangou, 2017, p. 428). Gebser's theory frames a transition into integral or holistic awareness according to our capacity to "leap" out of present certainties and "lean into the potential for a more sympathetic presence with the world" (Johnson, 2019, loc. 1785). I will now outline insights gained through the use of this theory in my PhD, and then my continuing exploration of its potential relevance to higher education.

Early educational research

I used Gebser's theory as an analytical lens in my doctoral research exploring the contemporary contribution of Froebel's holistic early years pedagogy. Froebel's ideas about education are underpinned by a Philosophy of Unity, through which "everything and every being … comes to be known only as it is connected with the opposite of its kind" (Froebel, 1887, p. 42). This is a principle that runs through Froebelian pedagogy and recognises scope for mutual learning in child and adult world relations. Many aspects of the current understanding of good early years practice retain continuities with Froebel's original pedagogy, but these practices have become separated from their originating holistic rationale. My research focus was the development of practice to address concerns about children's loss of nature contact within a global urbanisation process (Louv, 2005, WHO, 2023). I drew on Froebelian philosophy as it aligned with my practice-based observations of children's influence on adults in a natural environment in which I noticed children's potential to draw adults into more sensory and playful engagements with their surroundings. I wondered about the contribution of this perspective to the contemporary educational context with its concerns to connect children and nature and a growing urgency to address environmental impacts from a culturally conceived human separation from nature (Zylstra et al., 2014). Froebel outlined a "spherical law" (Liebschner, 1992) through which learning relations are multi-directional and continuous, and adults are called to "live with our children" (Froebel, 1887, p. 89). This potentially re-orients us to the multi-directional, or "networked relations," that the contemporary global crisis requires (Capra & Luisi, 2014).

My research explored the role of early education as a connective hub between children, families, communities, and the environment. Within this, I wanted to understand the potential for practice to address a complex combination of opportunity and orientation-related barriers to children's nature contact (Soga et al., 2018). There is growing recognition of the value of nature-connective experiences as healthy foundations for human and environmental health (Pretty et al., 2009). This research focused on a suburban pre-school that organised family trips to local natural environments. Data was gathered through children's wearing of video cameras during trips using a sensory ethnography approach (Pink, 2009). The video recorders on a chest harness offered an intimate capturing of children's embodied and verbal expressions and, through this, a perspective that was more physically immersive and less filtered by pre-existing knowledge. The footage formed the basis for a sensory elicitation interview with parents and offered them the opportunity to have "eyes to see, ears to hear and feeling to feel with the child" (Froebel, 1887, p. 73). This was significant, as Froebel recognised that adult perceptions become "dulled" towards their surroundings through an overemphasis on the written word and abstract knowledge. The sensory ethnography approach offered adults the opportunity through child-led perspectives to re-experience being "close to the ground and up against the full sensory qualities of things … that adult height and habits will later remove them from" (Chawla, 2002, loc. 2635). The time and space offered by interviews enabled parents to gain insight into the relational richness experienced by children and led them to reflect on the role played by such activity in their daily family lives.

I drew on Gebser's theory to analyse the child and adult data after being inspired by prior research in which it was described as offering a "vocabulary to talk about otherwise difficult-to-acknowledge aspects of children's experience of the natural world" (Chawla, 2002, loc. 2602). In this way, this theory provided a way of acknowledging and giving equal status to child and adult relations with the world. Use of Gebser's theory as an analytical lens facilitated attendance to the breadth of being and becoming expressed by adult and child (James & James, 2012). Research insights will be shared here to bring some of these difficult-to-express ideas to life.

This first extract illustrates the way seeing the world through the eyes of children can reveal the potential for a paradigm shift to a holistic perspective. Such a shift would reframe embodied place-making processes in terms of a capacity to be "exquisitely sensitive to context" (Leithwood et al., 1999, p. 4 cited in Bottery, 2016, p. viii) within a retuned awareness to life's interdependence. One parent's comment captured the perspective she gained through reflecting on the challenge of walking along a slippery, muddy path seen from her three-year-old daughter's point-of-view. She reflected on their collective encounter with the spontaneity of the natural world (Izenstark & Ebata, 2016) with an exclamation that such simple experiences are "just there, it's just there for you. The more you get out and experience, the more they want to do it again." Another parent described the way her three-year-old daughter had spotted nests in a tree whilst walking and asked questions which had prompted her to "start doing some learning myself." This mum then reflected on her own urban childhood and expressed an appreciation for the opportunity now to share learning about the world with her children.

> The world was just there, and I was trying to find my fit – until I had kids. Everything I do is new, and I get to share it with my own children. I'm just as excited because I haven't seen it before – but then I've got to try to be the grown up one, whereas sometimes I feel more like one of the kids. I didn't

really do it when I was younger, so for me, it's an even better experience to have that now. I'm learning nature now with them.

The research data also illuminated the way all families drew on familiar stories from home as a shared language in forming relations with life encountered in their wider local context (offering expressions of Gebser's mythic consciousness). A cross-referencing of this data with a psychoanalytic perspective on fairy tales (Bettelheim, 1976) showed a contrast in child and adult relational motivations. Examples of child-led imaginative storying included a three-year-old boy's repeated search for a big bad wolf or another who played with the idea of running away as a gingerbread man. Such examples demonstrate a child's drive to encounter risk and build confidence in their own capacities to do so. By contrast, the adult-led storying demonstrated an instinct to present the world as a safe place by drawing parallels between life encountered and the familiarity and safety of home. One Mum reflected on using stories as an activity characterised by sharing an experience of the world with her child. She explained "That's my way of relating it to something he knows. It's like a reference world to spike his interest. It's not something you do unless you're with children. Adults know the world."

Gebser emphasises the value of different ways of "knowing" the world and the need to integrate all relational experiences. The potential for this was observed in shared experiences of the natural world in which adult attention could be led by children's fascinations and refocused through this on aspects of surroundings that they had become accustomed to and stopped noticing over time. One parent commented on feeding ducks with her three-year-old daughter, whose attention had led her to "actually noticing the different ducks' behaviour, and you can see them underwater and how they feed." She also commented on the value in reviewing this from her daughter's perspective by stating, "It's nothing for us … yeah, it's a leaf, but that leaf, that daisy, that one feather, twig … massive. I want to build on that I think, would be a lovely thing." These examples illustrate the scope for adults to be drawn into experiences with young children that are integrative. There is alignment here with the contemporary understanding of pathways to nature connection – through contact, meaning, emotion, beauty, and compassion (Lumber et al., 2017) – although the way in which children can draw adults into such experiences is less well understood, as is the role of play in building connective relations. Froebel outlined relations between adult and child in which "Play truly recognised and rightly fostered, unites the germinating life of the child attentively with the ripe life of experiences of the adult and thus fosters the one through the other" (Froebel, cited by Liebschner, 1992, p. 24). The benefits of play are outlined in equally holistic terms and the potential to experience this as "joy, freedom, contentment, inner and outer rest, peace with the world" (Froebel, 1887, p. 55).

Drawing on Gebser's holistic theory has led to a scholarly assertion that new "lines of flight" are needed (Ingold, 2008). My exploration of a holistic perspective suggests that such "lines of flight" might lie within the fabric of our given conditions, and that adults might be prompted to re-engage with these through sharing in children's "everyday adventures and ordinary magic" (Gill, 2012). My research highlighted the pedagogical value of access to natural environments and materials, the power of our own creative capacities and a deep-seated need for archetypal stories. I am continuing to explore the relevance and contribution of these relational qualities as I now progress into higher education and work with future early years professionals.

Building bridges between pedagogies

I am currently working within an interdisciplinary team on an undergraduate community-development module in the Faculty of Arts, Humanities and Education. The module engages with multiple perspectives on community and is exploring this through a needs assessment of the Canterbury Christ Church University campus. Students have been invited to contribute their views in determining a priority to address, and data has been collected on an interactive map through analytical, and creative processes. I shared my PhD research as an example of a community-development initiative that engages with a holistic consideration of the child, family, community, environment, and society (The Froebel Trust, 2023). I have then drawn on the notion of the commons (Bauwens & Kostakis, 2014) to explore the application of a holistic perspective to a higher education context. A "commons" perspective considers resources in collective terms shared through relations "with materials, with tools and technologies, with human and non-human beings" (Kavedžija, 2022). Such a perspective frames health and well-being as occurring, not within the individual, but emergent through relationships, spaces, and communities, as "a commons that we must cultivate together" (Kavedžija, 2022). We explored this on our campus by mapping the relationships, spaces, and communities where we have felt and experienced conviviality and care. This included places where we find support or connection with human and more-than-human life, spaces to experience a sense of self, and for self-expression. The exercise facilitated students in hearing each other's experiences and learning about aspects of the campus of which they had previously been unaware. The discussion touched on the role of values, our relational responsibilities, and gave rise to identification of needs. Mapping such experiences made these conscious, collective, and offered the potential for their integration with physical locations and institutional structures and processes. It offered a means of conducting a more authentic assessment of student experience.

Continuing research in holistic education

I draw inspiration from these experiences, not only in current higher education teaching, but also in research exploring the contemporary benefits of holistic pedagogy. In one project I have used eco-linguistics (Fill & Penz, 2018) to analyse "The Education of Man" (Froebel, 1887) as inspiration in shaping an early years pedagogical response to the global environmental crisis. Froebelian pedagogy attends to "the centre and fulcrum of the self" (Froebel, 1887, p. 21), the "silent teaching of nature" (Froebel, 1887, p. 8), and promotes their continuity and connection. As educators, we are called to draw inspiration from the natural world and to give "'space and time' for children's development" (Froebel, 1887, p. 8). The significance of relational pedagogy is foregrounded. Reconnecting early education with its Froebelian historical roots can promote broader awareness of the contemporary relevance of a holistic educational approach. Such pedagogy promotes experience of natural environments, enables play responses, and embodies acts of care in supporting the flourishing of human and non-human life (Kemp, Josephidou & Watts, 2022).

In another project, I drew inspiration from early years pedagogical experience in exploring the role played by church toddler groups in supporting the spiritual flourishing of young children (NICER, 2022). Spiritual flourishing is understood as "a dynamic state of being, seen in nurturing the right relationship with self, others, creation, and the transcendent" (Casson et al., 2023), and has been linked theologically with the Gospel of John and reference made to living life to the full (John 10:10). The project set out to explore the significance of community experiences for supporting relational connections and a sense of belonging important to spiritual life. The pilot study investigated research methods to capture relational experiences between children, parents, church community, and environment. I drew on my early years background in shaping resources to support shared child and adult experience, and perspectives such as wellbeing and involvement scales (Laevers, 2005) and schematic play (Piaget, 1971) to aid data interpretation. Such work highlights the value in interdisciplinary research that can traverse world views and consider the connective potential of daily practices. The scope for benefits are multi-directional and early years pedagogy can contribute to the community support offered by church toddler groups, whilst gaining insight into the potential for play to support spiritual flourishing when understood as a life well-lived (Casson et al., 2023).

The challenge in listening to the self

The challenge of a holistic world view lies in recognising the self, our capacity to be reflexive, and acknowledge our own role in contributing to change. Do we uphold values that serve us individually or can we lean into a vision of the whole through institutional structures and processes? Can we trust the value in wisdom, commit to a "big picture" perspective, and the potential for evolution through self-distancing and cognitive integration (Li et al., 2020)?

For me the opportunity to engage with these challenging questions began when my partner and I, in our early twenties, moved to the west of Ireland and quickly fell in love with a mountainous terrain, extreme weather, and lyrical culture. We experienced a new sense of restfulness in our physical beings that contrasted sharply with our prior negotiation of the large-scale social systems of Manchester. Despite practical challenges, we enjoyed an easy transition into this small-scale community life. A shift in the balance was to occur, however, and that was through a change in our personal circumstances on starting a family. The precarious employment and lack of affordable housing in this small-scale rural economy made managing our new responsibilities a less easily surmountable challenge. We found ourselves making a highly unanticipated decision, and that was in returning to England and swapping the beauty in our surroundings with the steady convenience of a suburban life. We felt an acute sense of loss in this process as we sensed so much goodness for our young children in having freedom at a very young age to explore wild spaces from outside the door. However, we came to realise that their wellbeing was tied up with our own as a family, and that we needed to transition from a beauty in freedom to that in social support and connection.

I managed to find what I felt to be one compensatory measure in the move, and that was in the access it would offer the children to Steiner education. I had encountered a kindergarten in another Irish county and wondered if its focus on natural beauty and imaginative play offered a means to hang onto a taste of the wild, which they would lose in the environment through education. What I was to discover, however, was a path not only for the children but also myself as I followed them into the kindergarten and a training in its anthroposophical approach. I discovered an education that I felt to be exquisitely sensitive, not only to environment, but also to young children's growing needs. It is this that has since compelled me on a research journey in exploring the holistic educational paradigm and a commitment to this approach in higher education. However, this practice is as deeply rooted in the experience of living beside a mountain and witnessing my children's relations as it is enhanced by theories, which I now explore in seeking to understand it. I continue now in drawing on my biographical and professional experiences with children, families, communities, and environments in developing higher education practice.

Following a sense of beauty

My own experience leads me to question whether there is the potential to follow a sense of beauty in learning to belong to the whole. A sense of beauty is understood in nature connective terms as the aesthetic appeal of shape, colour, and form that can be found and appreciated in the natural world (Lumber et al., 2017). However, we might take inspiration from a holistic perspective in considering a conception of beauty in broader terms. St Augustine outlined beauty as the harmonious cooperation between parts of a whole, and Pseudo-Dionysis similarly outlined beauty as "the cause, the substance, the principle, the pattern, the measure and the purpose of all relationships" (Tatarkiewicz, 1978: p. 46). Thereby, there is a sense of beauty involving a pattern to which all elements find their fit, and Alberti (1492) asserted that for a "whole, there is only one right solution which is to harmonise and optimise the components" (cited by Tatarkiewicz, 1978: p. 135). In this way, a beautiful logic can underpin both the aesthetic appeal of surrounding life and an inner sense of satisfaction in the human experience of connection. According to Democritus, it is the conditions of life itself that compel us towards beauty, and in this we are guided both by the physical world and "an inner, metaphysical component" (Lehene, 2020, p. 145). Such connections lead me towards a continuing questioning about how we might listen to the environment ourselves and enable their interconnection, and to consider which voices are missing that we might now need to attend to, and how might we achieve balance when it is ongoing and dynamic.

Conclusion

I regard the value in a holistic paradigm as lying in its support for a collective wellbeing that can reinforce scope for a connective resilience. In this way, the value in conceiving the university as "an adaptive, innovating institution engaged in a continual co-evolutionary learning process with community and society" (Sterling, 2021, p. 5). I see the importance in this of attending to social and environmental relations that might enable our fullest capacities for meaning-based connections. Such relations might integrate the connective potential in immediate experiences and comprehension of their place within a whole, global, context. My perception is that it is equally important to acknowledge a sense of self at the centre, and for this to be supported in accessing experience and authentic expression. The holistic paradigm involves the means for a metaphorical shift from a linear to a "spherical" vision that offers multiple directions to be explored in seeking an ongoing healthy balance. It is this holistic metaphor that might now guide us in recognising ourselves as whole beings in whole relations through which we might learn to belong to the whole.

References

Alberti, L.B. (1492) L'Architettura (de re aedJ5catoria), ed. with Italian translation by Orlandi, 2 vols., Milan (1966).

Bauwens, M. and Kostakis, V. (2014) From the communism of capital to capital for the commons: Towards an open co-operativism. *Triple C: Communication, Capitalism & Critique. Open Access Journal for a Global Sustainable Information Society*, 12(1), pp. 356–361.

Bettelheim, B. (1976) *The Uses of Enchantment: The Meaning and Importance of Fairy Tales*. London: Penguin Press.

Bottery, M. (2016) Not so simple: The threats to leadership sustainability. *Management in Education*, 30(6), p. 97–101.

Braidotti, R. (2013) *The Posthuman*, Cambridge, UK and Malden, MA: Polity Press.

Capra, F., & Luisi, P.L. (2014). *The Systems View of Life: A Unifying Vision*. Cambridge University Press.

Casson, A.; Woolley, M.; Pittaway, A.; Watts, T.; Kemp, N.; Bowie, R.A.; Clemmey, K.; Aantjes, R. Paying Attention to the Spiritual Flourishing of Young Children in

Church Toddler Groups: A Scoping Study Evaluating the Feasibility of a Research Study in This Context. *Religions* 2023, *14*, 236. https://doi.org/10.3390/rel14020236

Chawla, L. (2002) Spots of Time Manifold Ways of Being in Nature in Childhood in Kahn, P & Kellert, S (eds.) *Children and Nature: Psychological, Sociocultural, and Evolutionary Investigations.* Cambridge, MA: MIT Press.

Davis, J. & Elliott, S. (2014) Research in Early Childhood Education for Sustainability: International perspectives and provocations. Oxford: Routledge Press.

Fill, A. & Penz, H. (2018) *The Routledge Handbook of Eco linguistics,* New York/Oxon, Taylor & Francis.

Froebel, F. (1887) *The Education of Man* (Translated by W. Hailmann). Dover: New York.

Froebel Trust (2023) Available at: www.froebel.org.uk/ (Accessed on: 22.7.22).

Gebser, J. (1949) *The Ever-Present Origin*, Athens, O: Ohio University Press.

Gill, T. (2012) *Celebrating ordinary magic and everyday adventures.* Available at: https://rethinkingchildhood.com/2012/04/26/everyday-adventures/ (Accessed on 12.10.21).

Ingold, T. (2008) "Bindings against Boundaries: Entanglements of Life in an Open World," *Economy and Space,* 40(8), pp. 1796–1810.

Izenstark, D. & Ebata, A.T. (2016) "Theorizing Family-Based Nature Activities and Family Functioning: The Integration of Attention Restoration Theory with a Family Routines and Rituals Perspective," *Journal of Family Theory & Review,* 8, pp. 137–153.

James, A. and James, A. (2012) *Key Concepts in Childhood Studies.* London: Sage.

Johnson, J. (2019) *Seeing Through the World.* Seattle: Revelore Press.

Kavedžija, I. (2022) Wellbeing: how living well together works for the common good, Rethinking Poverty. www.rethinkingpoverty.org.uk/wellbeing/wellbeing-how-living-well-together-works-for-the-common-good/

Kemp, N., Josephidou, J. & Watts, T. (2022) Developing an ECEC response to the global environmental crisis: The potential of the Froebelian-inspired "NENE Pedagogy." Available at: www.froebel.org.uk/uploads/documents/FT-NENE-Pedagogy-report-Dec-2022.pdf

Laevers, F. (2005) *Well-being and involvement in care settings. A process-oriented self-evaluation instrument.* Kind & Gezin and Research Centre for Experiential Education. www.kindengezin.be/img/sics-ziko-manual.pdf [Google Scholar].

Lehene, I. (2020) A Reconsideration on the Theory of Beauty: Selected Definitions, Concepts and Views on the Topic (Part I), *Periodica Polytechnica Architecture*, 51(2), pp. 209–219.

Li, K., Wang, F., Wang, Z., Shi, J., Xiong, M.. (2020) "A polycultural theory of wisdom based on Habermas's worldview," *Culture & Psychology,* 26(2), pp. 253–273.

Liebschner, J. (1992) *A Child's Work: Freedom and Guidance in Froebel's Educational Theory and Practice.* Cambridge: The Lutterworth Press.

Leithwood, K., Jantzi, D., and Steinbach, R. (1999) *Changing leadership for changing times.* Buckingham, UK: Open University Press.

Louv, R. (2005) *The Last Child in the Woods.* New York: Algonquin Books.

Lumber, R., Richardson M., Sheffield, D. (2017) "Beyond Knowing Nature: Contact, Emotion, Compassion, Meaning, and Beauty are Pathways to Nature Connection." PLoS *ONE,* 12(5), pp. 1–24. e0177186. https://doi.org/10.1371/journal.pone.0177186

Mahmoudi, S., Jafari, E., and Liaghatdar, M. (2012) "Holistic Education: An Approach for 21st Century," *International Education Studies,* Vol 5 (2), pp. 178 – 186.

Malone, K. (2018) *Children of the Anthropocene.* Gorleston: Palgrave MacMillan.

Meadows, D.H. (1999) Leverage points: places to intervene in a system, the Donella Meadows Project. Acad. Syst. Change. Available online at: https:// donellameadows.org/archives/leverage-points-places-to-intervene-in-asystem/

National Institute for Christian Educational Research (NICER) (2022) Nexus Research at Church Toddler Groups. Available at: https://nicer.org.uk/toddlers-in-the-nexus

Piaget, J. (1971) The theory of stages in cognitive development. In D. Green, M.P. Ford, & G. B. Flamer (eds.), Measurement and Piaget (pp. 1–11). New York: McGraw-Hill.

Pink, S. (2009) *Doing Sensory Ethnography*. London: Sage Publications.

Pretty, J., Angus, C., Bain, M., Barton, J., Gladwell, V., Hine, R., Pilgrim, S., Sandercock, S., and Sellens, M. (2009) *Nature, Childhood, Health, and Life Pathways. Interdisciplinary Centre for Environment and Society Occasional Paper 2009-02.* University of Essex, UK.

Soga, M., Takahiro Yamanoi, T., Tsuchiyaa, K., Koyanagi, T.F. and Kanaib, T.. (2018) "What are the Drivers of and Barriers to Children's Direct Experiences of Nature?" *Landscape and Urban Planning,* 180, pp. 114–120.

Sterling, S. (2021) "Concern, Conception, and Consequence: Re-thinking the Paradigm of Higher Education in Dangerous Times." *Frontiers in Sustainability*, 2:743806. doi: 10.3389/frsus.2021.743806

Taylor, A. and Pacini-Ketchabaw, V. (2015) "Learning with CChildren, Ants, and Worms in the Anthropocene: Towards a Common World Pedagogy of Multispecies Vulnerability," *Pedagogy, Culture & Society*, 23(4), pp. 507–529.

Tatarkiewicz, W. (1978) "Istoria esteticii" (The History of Aesthetics), Meridiane, vol. III, Bucharest.

World Health Organisation (2023) *Urban Health,* www.who.int/health-topics/urban-health#tab=tab_1

Yiangou, N. (2017) "Is a New Consciousness Emerging? Reflections on the Thought of Ibn 'Arabi and the Impact of an Integral Perspective," *World Futures*, 73(7), pp. 27–441.

Zylstra, M.J., Knight, A.T., Esler, K.J., & Le Grange, L. (2014) "Connectedness as a Core Conservation Concern: An Interdisciplinary Review of Theory and a Call for Practice," *Springer Science Reviews*, 2, pp. 119–143.

7 Developing sustainability education through small-scale interventions

Stephen Scoffham

Introduction

What is education for, and how can we best set about it? What do we need to do to equip students, children, and young people for the future? And what are the values and principles that underpin current practice? These are fundamental questions about education, which have exercised both philosophers and practitioners for centuries. How such questions are framed and the way they are answered depends to some considerable extent on the context and wider social and cultural factors. Today, as the global environment crisis gathers momentum, finding new ways to live within planetary limits has become an urgent imperative. This is impacting how education is conceived and conducted.

Education is necessarily orientated towards the future as it focuses on the needs of the younger generation who have their lives before them. However, predicting the future is at best an uncertain process, and past attempts have often proved dramatically wrong. Despite this, there are some aspects of our lives where there are enduring challenges that seem likely to persist in the years to come. The educational philosopher, Gert Biesta (2015), identifies three areas or questions he believes provide a broad frame of reference that permeates all our endeavours. These are:

(1) democracy and how to live together despite our differences,
(2) the care of others, particularly those who are not able to care for themselves and
(3) ecology and how to manage our collective lives on a finite planet.

What is striking about these questions is that whilst democracy and the care of others have a long history, living within planetary limits is a relatively new concern. Although Biesta doesn't develop this point directly, his argument suggests that educational practices need to be recalibrated to take account of, and to address, sustainability issues. Many young people would agree. The

DOI: 10.4324/9781003286516-11

climate strikes and youth protests which erupted around the world in 2019–2020 are just one indication of what Stephen Sterling calls their 'profound concern and fierce hope' for the future of the world (2022. p. 75). Environmental anxiety is deeply engrained in the psyche of many young people today.

Sustainability education

Sustainability education is a fluid concept that embraces a range of interdisciplinary approaches to environmental learning, including Education for Sustainable Development (ESD), Education for Sustainability (EfS) and Environment and Sustainability Education (ESE). It also overlaps with a range of educational initiatives, such as Forest Schools and Outdoor Learning. The ambiguity that surrounds this area of education is hardly surprising given that ideas about sustainability are themselves both evolving and contested. Early definitions of sustainability focused largely on ecology and the damage that pollution was causing to the natural environment. Since then, the idea that sustainability involves social and economic dimensions has steadily gained traction. This means that the way people lead their lives, both individually and collectively, is now seen as an essential component of a broader understanding of sustainability. There is also a growing recognition that environmental problems cannot be considered in isolation but interact in complex and often unexpected ways.

There is no simple definition of sustainability education that can capture these complexities. Biesta talks about how an education for the future involves moving from the 'ego-centric' to the 'eco-centric', by which he means a shift in mindset away from individual survival towards an awareness of planetary needs. We have to ask, he declares: 'What is desirable for the child, what is desirable for the life the child lives with others and the life we have together on a vulnerable planet with limited capacity' (2015, p, 8). The working definition adopted in this chapter sees sustainability education in similar terms but stresses interactions. This approach emphasises how a sustainability mindset involves reconnecting (a) to ourselves (b) to others and (c) to the environment at a range of scales. Figure 7.1 illustrates how this approach brings people and the natural world together in a dynamic and reciprocal network. Viewed from this perspective, sustainability education can be seen both as an invitation to reform the curriculum and as a challenge to address and adapt to current circumstances.

So how can sustainability education be developed in a university environment, where there are multiple stakeholders and conflicting priorities (Haddock Fraser et al. 2018)? A number of examples are presented below, which illustrate initiatives that have been undertaken at Canterbury Christ Church University. They range from personal development and team building to research into personal and professional motivation.

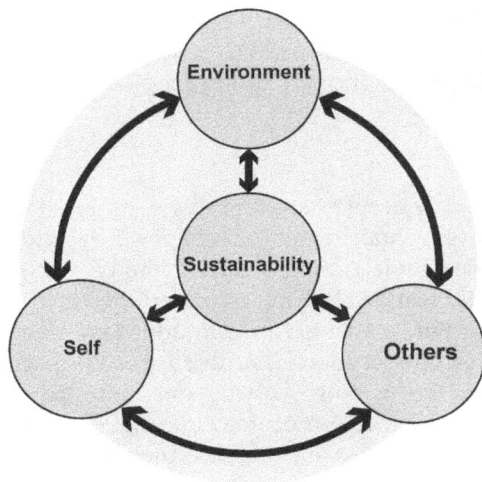

Figure 7.1 Interactions and connections are at the heart of a sustainability mindset.

Source: After Scoffham and Rawlinson, 2022.

Personal development – my own story

My own interest in the environment and sustainability is hard to isolate. Some people can point to eureka! moments, when all their ideas suddenly fell into place, but my, particular trajectory was much more gradual and evolved over time. In general terms, the opportunity to interact with the natural world and travel to different places in my youth was undoubtedly a significant factor. As a child I was lucky enough to have good access to the countryside and the freedom to explore it unaccompanied. Later, I cycled along local lanes, seeking out historic sites and points of interest. For many years I also went on annual walking holidays with my parents to the Italian Alps. As a student, I travelled independently, venturing overland to Turkey and India, immersing myself in different cultures and traditions.

These experiences had a significant influence on both my professional and my personal life. Having obtained a first degree in philosophy and history, I trained as a primary school teacher and began to engage with the literature around environmental issues and global inequalities. I was inspired by initiatives such the *Art and the Built Environment* (Adams and Ward, 1982) and the Schools' Council *Time, Place and Society* project. At about the same time, I was introduced to *World Studies 8–13* (Fisher and Hicks, 1985) and stumbled across the journal *New Internationalist* and the *State of the World Atlas* (Kidron and Segal, 1981) in my local Oxfam bookshop. However, it was only when I was appointed as a lecturer at CCCU that I began to engage with sustainability as an educational concept and learn about the theories and research that underpinned it.

I began by focusing on the headline news about environmental destruction. Sadly, there were plenty of examples to choose from, ranging from acid rain, desertification, the decimation of wildlife, and the melting of the ice caps on the one hand, to social issues such as global inequality and international debt on the other. However, before long I realised the complications involved in teaching about these things and the challenge of developing sufficient knowledge and expertise across so many different themes and disciplines. It was also apparent that simply providing students with facts and problems was liable to be counterproductive. Indeed, nothing seemed to provoke negative responses more effectively than Al Gore's film *An Inconvenient Truth* (2002), which highlighted environmental catastrophe clearly and firmly but left viewers feeling angry and helpless.

Research bears out my personal observations. In a fascinating study, Kollmuss and Agyeman (2012) found that social factors and individual circumstances are much more powerful than factual knowledge in developing pro-environmental behaviours. Subsequent research conducted by Moss et al. (2016) and Marcinowski and Reid (2019) reached a similar conclusion. Recognising that our behaviour is contextualised and embedded in a nexus of cultural, social, political, and historical factors means that appealing to the emotions and generating peer pressure is liable to be more effective in changing opinions than factual knowledge alone. In curriculum terms this suggests that the impact of the environment crisis needs to be communicated as much through the arts and humanities, as through the sciences. Meanwhile, on a personal level, the contradictions between what I was trying to do as an educator and the environmental footprint associated with my own lifestyle were also glaringly obvious. Whilst I was sincere in what I was doing in the classroom and convinced that it mattered, I was certainly not walking the talk.

Building a community of practice

It was clear from my own experience that individuals working alone are liable to have only a limited impact on courses that involve multiple tutors and stakeholders. Enhancing the capability and capacity of staff with respect to sustainability education is an institutional priority that was recognised at CCCU in 2012 with the launch of a staff development programme called the *Futures Initiative*. A number of principles underpinned the project from the outset, including a very open 'grassroots' approach to curriculum development, an inclusive culture that supported creativity and innovation, and the recognition that sustainability education involves different ways of knowing. This is encapsulated in the phrase 'head, hands, heart'. It was also understood that staff development is a long-term process, and that there could be no quick fixes. The overall aim was to build a network of colleagues, especially those who were already interested in the environment and sustainability issues, in the hope that they would encourage and inspire others (Scoffham, 2016).

A key component in the early years of the *Futures Initiative* was a residential programme at a peaceful old manor house and its associated organic dairy farm. The site, well away from the pressures and concerns of everyday academic life, provided an ideal 'third space' (Wasam-Ellam, 2010), where colleagues could explore emerging ideas in a safe, low-stakes setting. The chance to meet and share ideas with colleagues from other departments and faculties with similar interests was widely acknowledged and highly prized by many participants. But the intangible benefits were probably just as significant. One tutor commented 'It was like taking a deep, deep breath'. Another said, 'I am coming away with something which is difficult to talk about in hard language. It's almost like a hue, almost like a shade difference'.

The networks and collaborations that developed from these residential events exhibited many of the features of what (Wenger, 1998) terms a 'community of practice'. Here, colleagues develop a shared repertoire of experiences, stories, and tools as they pursue their common interests and develop their professional networks. What is particularly significant is that, over a decade later, many of the embryonic communities that resulted from these residential programmes have continued to flourish and evolve. This holds out the possibility of building a critical mass that will eventually stimulate more widespread institutional transformation.

Curriculum innovation

The chance to exhibit a major photographic exhibition called *Whole Earth?* at CCCU in 2015 opened up further possibilities. Consisting of six banners, each 10 metres long and framed by Bob Dylan's famous protest song *A Hard Rain's A Gonna Fall*, the exhibition highlighted a variety of environmental and sustainability issues together with some possible solutions. The text also contained panels with questions for both staff and students about how sustainability could be meaningfully incorporated into programmes and courses in their own curriculum areas. The overall aim of the exhibition was to encourage universities to review their curriculum practices and to focus on the most pressing issues of the age – how we can continue to live and prosper within the limits of the planet that supports us.

The visual arts are time-honoured methods of constructing meaning and communicating ideas that transcend language and often touch people at a deep level. Zajonc (1980) is one of a number of researchers who has found that exposure to images appears to have a greater impact on behaviour than abstract or technical information. Others, such as Iozzi (1989, p. 3) contend that the emotions are the 'gateway' to learning. This is supported by neurological evidence (Immordino-Yang and Damasio, 2007) that shows how emotional reactions to stimuli develop prior to cognition, and that emotions and value systems guide the way we process and assess information. If developing knowledge and understanding involves more than purely cognitive processes, then a hard-hitting photographic exhibition might be more effective and have more long-lasting impact than more formal presentations on environmental issues. At the very least it could be seen as a valuable complement.

The research and analysis conducted immediately after the *Whole Earth?* exhibition was dismantled certainly indicates that it stimulated a considerable range and diversity of responses (Scoffham, 2018). The exhibition generated discussion, made people think, and led them to question their values. Many tutors commented on the visual impact of the photographs, which 'hit home more than words'. One respondent (a pro vice chancellor) declared that it raised 'fundamental questions' about how they lived their life and enacted their role in the university. Meanwhile, one particularly committed student said that the exhibition had prompted her to set up an environmentally focussed cleaning business. In summary, the exhibition seemed to provide a narrative to which staff and students could respond in different ways, and it generated a particularly positive response in those who were already interested and engaged in environmental issues.

Meaning and motivation

One of the many advantages of working as a small and cohesive team at CCCU is that we have been able to have impromptu conversations about sustainability issues and forge new understandings together. A question that particularly caught our interest was what had drawn us to sustainability and why we had decided to focus our careers around it. We decided to explore further and devised a small-scale research project in which we engaged Victoria Field, a creative writer, to develop fictional accounts based on our professional lives. We hoped that this approach might, as Peter Clough puts it, provide a means to uncover 'those truths, which cannot otherwise be told' and open up 'a deeper view of life in familiar contexts' (2002, p, 8).

The seven accounts that resulted from the project acted as provocations that stimulated further conversations. Here is an extract from one of the stories based on an individual (who happens to be male) which focuses especially on meaning and motivation:

> This person has lived here all his life but he's also lived elsewhere all his life. He dwells in the whole wide world. For all we know he may have lived here in previous lives too and may live here again. As he walks through (the city), he unravels a thread of golden light, some can see it some cannot. He says its hard to say exactly what the thread is. It's had different names.
>
> So what is he doing, our city walker, our *flaneur*? Why does he wander among the streets and buildings? I'm turning over stones and looking round corners, he replies. Its what I've done since I was a child. I am looking for something called meaning.
>
> (Field, 2020 pp. 15–19 Abridged and adapted)

The way this narrative focuses on the search for meaning drills further into the well-springs of our behaviour. Human beings have an innate desire to make meaning from the various experiences that punctuate their lives. Our vision for ourselves and the future is an essential part of our identity and sustains our lives. The environment crisis requires us to construct new visions. As Jeremy Lent puts it in his extensive review of social and cultural history, 'We need a new way of finding meaning based on an underlying and all-infusing sense of our connectedness to ourselves, others and the natural world' (2017, p, 440). Exploring what this might be and how it can be expressed is a challenge that can be conducted in many forums and by all sections of society. In a memorable turn of phrase, Arran Stibbe suggests that in different ways we are all involved in the task of 're-writing and re-speaking the world' (2015, p, 193). This is surely what 'good education' is all about.

Small-scale interventions and clumsy solutions

These examples all illustrate different ways of raising sustainability awareness which, rather than simply relying on scientific evidence and the transmission of knowledge, explore alternative ways of knowing and thinking with staff and students. They provoked emotional responses and stimulated imaginative thinking amongst those involved. They are necessarily small scale because they engaged limited numbers of individuals located within a higher education institutional framework which is calibrated according to an accountability culture and neoliberal values. Over the years there have been many calls for a paradigm shift in education, but this is yet to happen, and it remains unclear how this might come about, especially given existing hierarchies and power structures. It is also important to recognise along with Biesta (2016) that education is necessarily a slow and uncertain process, and that its outcomes can never be guaranteed. In such circumstances, working around the edges of existing structures and developing capacity and understanding on multiple levels seems both timely and pragmatic.

The way that different initiatives can enhance each other and build a momentum for change is illustrated in the 'paradox model' developed by Kemp and Scoffham (2020). The model (see Figure 7.2) provides a navigational tool for sustainability education constructed around two challenges:

(a) how to develop authentic responses within the context of existing HE structures and processes, and
(b) how to reconcile the demand for immediate action with the more gradual processes of learning and education.

The model embraces these tensions and offers a framework that validates alternative responses. It reconciles both calls to resist current practices and approaches that favour alignment. The demand for immediate action is set alongside the slow processes of developing wisdom and judgement. The

The Paradox Model

Fast

School

Class

Pupils

Resistance ◄──────────────► Alignment

Slow

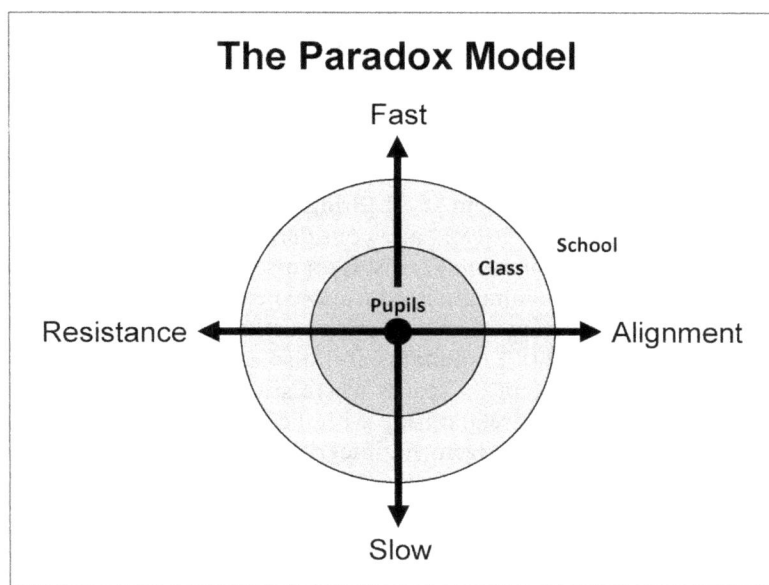

Figure 7.2 The paradox model provides a framework which locates different approaches within the tensions that permeate sustainability education.

Source: After Kemp and Scoffham, 2022.

model also provides a lens that operates from the level of the individual student or tutor to an institutional scale where a university may express its values and commitment in its strategic plan or mission statement. Recognising the validity of different responses to sustainability education creates resilience and flexibility, just as in the natural world a habitat gains strength from diversity. And the model also recognises the importance of context. No two places or environments are quite the same.

The current environmental crisis, like any emergency, plays into fears about an unknown and uncertain future. In education, as in other spheres of life, it is tempting to seek grand solutions that will solve all our problems once and for all. However, this is to misinterpret the nature of the challenges that lie ahead. Sustainability education is essentially a 'wicked problem' (Rittel and Webber, 1973) and has many dimensions that make it ill-defined, ambiguous, and even contradictory. Mike Hulme (2014), in his analysis of the arguments around climate change, contends that even the way a wicked problem is presented changes its characteristics. This leads him to favour multiple 'clumsy solutions' that have the potential to reframe issues and achieve small incremental gain. With respect to sustainability leadership, Mike Bottery (2016) takes a similar stance, arguing for a variety of approaches which, although they will never be perfect, are the best that can be done at the time. One-stop solutions, whilst they may seem initially appealing, are fraught with hubris, just as large-scale geo-engineering projects are tainted by megalomania.

There is a further point. The initiatives described above led a number of those involved to ask questions about how they lived their lives. This is important because the values and beliefs that guide our behaviour are often so deeply embedded that we simply take them for granted: indeed, we sometimes do not even recognise them at all. Bringing to the surface what Stibbe calls 'the stories we live by' (2015:3) and considering whether they are appropriate in the current environmental crisis, is an essential part of developing a new mindset. This is a fundamentally educational endeavour, as new learning often involves deconstructing and rearranging existing ideas and assumptions. Furthermore, as Biesta (2017) maintains, it is also a disruptive and uncertain process that involves opening up spaces where students can encounter their freedom and establish a new relationship with their desires. Such an approach places an emphasis on good judgement, integrity, and moral wisdom – all key components of sustainability thinking.

Saving the planet

Sustainability is often presented as if it involves a set of actions to 'save the planet'. Saving the planet is a slogan that has wide appeal and is touted in schools with pupils from the earliest age. It also provides the backdrop for sustainability initiatives in higher education and university settings. The idea that any individual or particular group can actually 'save the planet' is clearly hollow. If they could, then they would have unimaginable power and unimaginable responsibility. The slogan also begs the question of what we are trying to save the planet for or from. In the 1970s and 1980s environmental issues were often presented as if they were problems that were somehow happening 'out there' irrespective of human activity. The idea that people are part of problem and that what we do is causing environmental destruction has now taken root. As Kumar and Howarth remind us in a simple but telling phrase, 'what we do to the Earth we do to ourselves' (2022, p. 16). This leads to the conclusion that we need to save ourselves from our own actions rather than somehow solve the difficulties of the planet, which can and will manage perfectly well without us.

But the idea of doing something to 'save the planet' does have validity on a different level. As they learn about what is happening to world and the damage that is being done to the beauties and treasures of nature, people naturally want to do something practical in response. Students, many of whom are at a point in their lives when idealism burns particularly brightly, are especially keen to make a difference. The actions we take to live more sustainable lives, however small, contribute to our sense of integrity and authenticity. That's why they matter. Even if global trends are going in the wrong direction, we can still do something. David Orr (2003) talks about the challenge of being part of a counter movement in an article entitled 'Walking North on a Southbound Train'. The point is that, even if we are deeply enmeshed in current neoliberal social and economic trends and lead lives that have large ecological footprints, there are still plenty of small actions we can take that will help to shift the current direction of travel. This also applies at an institutional level, where tutors in universities also have scope to go against the flow within their own spheres of influence.

Underpinning the call for action to save the planet is the vision of a better world and belief in a brighter future. Without hope, people are liable to enter a state of denial or give up altogether and, whilst we need to be realistic about what might be achieved, we are sustained by our ambitions for the future. Macy and Johnstone see this as a profound psychological truth. They argue that that by harnessing hope to galvanise action we can all make a contribution to healing the world. This then opens up the door to an 'amazing journey', which strengthens and deepens our sense of being alive (2012, p. 2). In the current circumstances, restoring a sense of purpose and meaning in the face of disorientating challenges is a high priority. It is a task that education is well placed to undertake.

Conclusion

We are all compromised by the environment crisis, as it requires us to live our lives in different ways and to step outside current social norms. However, it is now clear that universities, along with society at large, are going to have to make radical changes if they are to adjust to the challenges that lie ahead. How might these changes come about? Rather than focussing exclusively on grand solutions, small-scale interventions such as the ones described in this chapter offer a pragmatic approach that can be implemented immediately by individuals and small groups in HE settings. Additionally, they involve different ways of knowing and being, which have the potential to raise questions about values, uncover assumptions, and provoke changes that are attuned to specific needs and circumstances.

The sustainability agenda involves reconnecting with ourselves every bit as much as it involves reconnecting with the world around us. It involves, as Pope Francis declares in *Laudate Si*, integrating 'questions of justice in debates on the environment, so as to hear both the cry of the earth and the cry of the poor' (2015 para. 49). Focussing on pedagogies of disruption and reconnection has the potential to initiate a shift in thinking which, if multiplied, could build into a critical mass for change. Some twenty years ago the Earth Charter (2000:1) invited people to 'join together to bring forth a sustainable global society founded on respect for nature, universal human rights, economic justice and a culture of peace'. There are lots of different ways those of us who work in higher education can help to bring this about, and we can all contribute to the process of developing an education that matches the needs of the future.

References

Adams, E. and Ward, C. (1982) *Art and the Built Environment*. London: Longman.

Biesta, G. (2015) 'The Duty to Resist: Redefining the Basics for Today's Schools', in *Research in Steiner Education*, 6: 1–11.

Biesta, G. (2016) *The Beautiful Risk of Education*. London: Routledge.

Biesta, G. (2017) *The Rediscovery of Teaching*. London: Routledge.

Bottery, M. (2016) *Educational Leadership for a More Sustainable World*. London: Bloomsbury Academic.

Clough, P. (2002) *Narratives and Fictions in Educational Research*. Maidenhead: Open University Press.

Earth Charter (2000) available at https://earthcharter.org/library/the-earth-charter-text/

Field, V. (2020) *Not Utopia … But Maybe*. Canterbury: Canterbury Christ Church University.

Fisher, S. and Hicks, D. (1985) *World Studies 8–14*. London: Longman.

H.H. Pope Francis (2015) *Laudato Si,* available at www.vatican.va/content/francesco/en/encyclicals/documents/papa-francesco_20150524_enciclica-laudato-si.html

Haddock Fraser, J., Rands, P. and Scoffham, S. (2018) *Leadership for Sustainability in Higher Education*. London: Bloomsbury Academic.

Hulme, M. (2014) *Can Science Fix Climate Change?* Cambridge: Polity Press.

Immordino-Yang, H. and Damasio, A. (2007) 'We Feel Therefore We Learn: The Relevance of Affective and Social Neuroscience to Education' in *Mind, Brain and Education*, vol 1.

Iozzi, I.A. (1989) 'What Research Says to the Educator: Part One: Environmental Education and the Affective Domain', in *Journal of Education for Sustainable Development,* 20(3), 3–9.

Kemp, N. and Scoffham, S. (2022) 'The Paradox Model: Towards a Conceptual Framework for Engaging with Sustainability in Higher Education', in *International Journal of Sustainability in Higher Education*, 23: 1.

Kidron, M. and Segal, R. (1981) *The State of the World Atlas*, London: Pluto Press.

Kollmuss, A. and Agyeman, J. (2012) 'Mind the Gap: Why Do People Act Environmentally and What are the Barriers to Pro-Environmental Behaviour?' in *Environmental Education Research*, 8(3), 241–259.

Kumar, S and Howarth, L. (eds.) (2022) *Regenerative Learning; Nurturing People and Caring for the Planet*, London: Global Resilience Publishing.

Lent, J. (2017) *The Patterning Instinct, A Cultural History of Humanity's Search for Meaning*. Amherst, NY: Prometheus Books.

Marcinkowski, T. and Reid, A. (2019) 'Reviews of Research on the Attitude-Behaviour Relationship and Their Implications for Future Environmental Research' in *Environmental Education Research* 25(4), 459–471.

Moss, A., Jensen, E., and Gusset, M. (2016) 'Probing the Link Between Biodiversity-Related Knowledge and Self-Reported Proconservation Behaviour in a Global Survey of Zoo Visitors' in *Conservation Letters*, 10(1), 33–40.

Macy, J. and Johnstone, C. (2012) *Active Hope*. Novato, CA: New World Library.

Orr, D. (2003) 'Walking North on a Southbound Train', in *Conservation Biology* 17(2), 348–351.

Rittel, M. and Webber, H. (1973) 'Dilemmas in General Theory of Planning', in *Policy Sciences* 4(2), 155–169.

Scoffham, S. (2016) 'Grass Roots and Green Shoots: Building ESD Capacity at a UK University', in Davim, J.P. and W. Leal Filho (eds.). *Challenges in Higher Education for Sustainability*, Cham, Switzerland: Springer International.

Scoffham, S. (2018) '"*Whole Earth?*" Using an Exhibition to Raise Sustainability Awareness at a UK University', in *Journal of Education for Sustainable Development* 12: 2 160–175.

Scoffham, S. and Rawlinson, S. (2022) *Sustainability Education: A Classroom Guide.* London: Bloomsbury Academic.

Stibbe, A. (2015) *Ecolinguistics: Language, Ecology and the Stories We Live By*. London: Routledge.

Sterling, S. (2022) 'Profound Concern, Fierce Hope', in Kumar, S., and Howarth, L. (eds.) (2022) *Regenerative Learning; Nurturing People and Caring for the Planet.* London: Global Resilience Publishing.

Wasam-Ellam, L. (2010) Children's Literature as Springboard to Place-Based Learning', in *Environmental Education Research*, 16 (3–4), 279–294.

Wenger, E. (1998) *Communities of Practice: Learning, Meaning and Identity*. Cambridge: Cambridge University Press.

Zajonc, R.B. (1980) 'Feeling and Thinking: Preferences Need no Inferences' in *American Psychologist*, 35(2), 152–175.

8 Walking towards embodied understanding

Sonia Overall

Introduction

This chapter is situated in relation to my creative practice and teaching as a walking writer and psychogeographer. It draws on my use of attentive walking methods to encourage immersive experiences of place, as a tutor in higher education and in broader public contexts, such as walks, workshops, and residencies. Through the discussion that follows I will bring together the psychogeographical attitude of attentive walking, the concepts of 'enduring time' (Baraitser, 2017) and 'pedagogies of attention' (Clarke and Witt, 2017), with a hyperlocalised field of practice, to propose a model for attending to place that is sustainable and embodied.

Why walking?

Beyond the purely functional or essential, walking can be playful, creative and provocative. Walking facilitates sensory experiences at large, giving us access to details of place, including textures, smells, sounds, and associated emotions. It is an essentially embodied experience, connecting us to seasonality and shifting environmental conditions through the effects of weather: walking is subject to constant variation, from the ground we tread on to the air we breathe. Walking is, as Tim Ingold (2010) frames it, a form of embodied perception that adjusts to shifting factors of terrain and weather. 'Walking ... is not the behavioural output of a mind encased within a pedestrian body. It is rather, in itself, a way of thinking' (Ingold, 2010, p. 135). Walking can be adapted to suit a variety of modes and bodies, as the diversity of papers and provocations in the collection *Walking Bodies* (Billinghurst et al, 2020) attests.

Walking also has the potential to foster new connections between people, place, and the nonhuman. An Arts and Humanities Research Council funded research project into the role of walking during lockdown, *Walking Publics/Walking Arts* (Rose et al, 2022), considers the relationship between walking, creativity, and wellbeing. The project's report findings make an excellent case for the restorative and re-enchanting potentialities of walking:

DOI: 10.4324/9781003286516-12

Walking can be a powerful tool, bringing joy, delight and comfort. It can also facilitate an enhanced sense of community connection and belonging, both in local neighbourhoods and with nature and green space. Simple creative walking initiatives can enhance all these benefits … During the pandemic, small, everyday moments of magic were created by our footsteps. We believe these desire lines can develop into sustainable and important new paths.

(Rose et al, 2022, p. 3)

What is 'attentive walking'?

I first define 'attentive walking' in my article 'Walking against the current' as a method of walking 'without agenda' (Overall, 2015, p. 14) that differs from *flâneurie*. Attentive walking makes use of the psychogeographical practice of the *dérive* or drift, which encourages the walker to follow their curiosity rather than a defined direction or purpose. However, attentive walking necessitates an acuteness of interest, a willingness to submit to what the environment offers the walker; to tune in to what it presents. This is distinct from the casual, strolling practice of the *flâneur* or *flâneuse*, with its sense of ambivalence or entitlement; and from the Situationist position of psychogeography as an essentially urban and primarily political act, as found in the work of key practitioners like Debord (Knabb, 1981). Attentive walking can in itself be political and radical, challenging human behaviours and our expectations of place in surprising ways.

Attentive walking encourages a sharpened focus, making use of the heightened awareness that walking brings. It amplifies our attention to the sensory, physiological, psychological, and aesthetic experiences of place. This method lends itself to speculative and site-specific responsive writing in my own practice, and is readily transferrable to creative writing exercises with students. In a wider context, it can be used to connect more deeply with place, and to enrich our understanding of it. Attentive walking can lead to the generation of outputs or creation of ideas, but is also valuable as a purely experiential practice.

Using scores and prompts: attentive and 'attitudinal walking'

Adding a layer or lens to attentive walking gives the practice additional focus. I have defined this as 'attitudinal walking', a form of attentive walking practice 'employing one or more conscious intentions or attitudes' (Overall, 2019). These 'attitudes' are parameters for the walk and can take the form of a walking score or prompt, adding an element of 'questing' or providing a specific frame – sensory, narrative or conceptual – through which to explore and experience the environment. Taking a question, concern or idea for a walk becomes attitudinal when the walker seeks answers in the environment.

Constraints, instructions, invitations, and provocations can be applied to an attentive walk to disrupt everyday ways of seeing and sensing place. Examples of attitudinal prompts can be found in my book *Walk Write (Repeat)* (Overall, 2021), and online in 'Pick a Drift' (2020), offering DIY walking scores for the 2020 Fourth World Congress of Psychogeography.

The application of an attitudinal lens can create interest in places and spaces that may have become stale through overfamiliarity, as experienced with enforced hyperlocalised walking. I will return to this later in the chapter but, here, an attitudinal walking project I initially generated in response to the Covid-19 pandemic, 'Distance Drift', offers a useful recent example. 'Distance Drift' is a remote, participatory project hosted on Twitter. I began the project in the first UK lockdown in April 2020, sharing walking scores or prompts designed to be suitable for short local walks outdoors, and for use indoors by those shielding, or recovering, from Covid. I release a score at a set time each week; this can be responded to synchronously, with an hour-long walk, or asynchronously after the event. Participants follow the hashtag #DistanceDrift to receive the score, post their responses to it, and engage with other walkers. At the time of writing (March 2023), 'Distance Drift' has continued to meet online every Sunday since its inception. There are, therefore, many attitudinal prompts in the #DistanceDrift archive available to try, from colour walks and sensory mapping to celebrating calendar customs and applying motifs from literature and pop culture as processes of defamiliarisation. By way of example, this score during a spell of hot weather was deliberately playful, encouraging walkers to apply a holiday lens to a sensory exploration of the environment:

> Morning! It's another hot one in the UK, so for today's #DistanceDrift, we're going out for ice cream. As you wander, indoors or out, gather scoops of ice-cream colours, scents, shapes & words; seek suitable toppings or sauces; grab a cone or a cup & spoon. Flakes optional.
>
> (Overall, 2022)

'Distance Drift' introduced me to the work of two researchers in Geography subject pedagogy. Dr Helen Clarke and Dr Sharon Witt have become regular contributors to the Twitter thread, and through the window of their playful posts, detailing attentive encounters and interactions with place, I recognised a kinship in our thinking. Clarke and Witt define their Geography fieldwork methods as 'pedagogies of attention' (Clarke and Witt, 2017). They position their approach 'with a call for a paradigm shift from learning *about* to learning *with* the world (Clarke and Witt, forthcoming, p. 2). Their practice is essentially embodied, experiential, and rooted in place. In common with my own psychogeographical practice, they advocate a form of fieldwork that attends to the everyday as a source of re-enchantment:

> Field-visiting is a different sort of practice – a thinking with other beings, animate and inanimate, and a propensity to find everything, especially the

mundane and everyday, fascinating and thought provoking. Field-visiting is an entanglement (p. 14).

Attending to spaces in these ways enables creative responses to and re-imaginings of place. These processes are suitable for application to learning and teaching activities, practice research methods, or simply for personal exploration and immersion. By encouraging the use of the senses, attentive practices foster embodied experiences and facilitate tactile and haptic encounters. They are immersive and defamiliarising, revealing elements of place that we might otherwise overlook. It is this looking again, this experiencing afresh, that can lead to a re-enchantment with place.

Looking away

Evidence of climate change is becoming increasingly tangible even in the temperate climate of the UK, and the effects of global heating more wide-ranging. Awareness of environmental change and the prospect of climate catastrophe inevitably result in adverse mental health and wellbeing. A paper by Hannah Comtesse and fellow researchers summarises the issue:

> The mental health effects of climate change-related environmental changes may occur directly due to natural disasters and extreme weather events (e.g., as traumatic stress reactions due to floods or wildfires) or indirectly through long-term, mostly secondary stressors that foster mental disorders (e.g., less secure food supply, problematic process of reconstruction of housing or infrastructure, forced migration due to increasing temperatures). Moreover, the impacts of climate change on mental health range from acute (e.g., distress during or after floods or heatwaves) to chronic (e.g., distress due to permanent landscape changes after tornados or mountains' loss of snow cover).
> (Comtesse et al, 2021, p. 1)

While these acute and chronic situations are already affecting many around the globe, those of us in temperate, developed countries are still physically removed from most extreme weather events and environmental disasters. Comtesse et al. go on to identify ecological grief as 'a natural response to ecological loss, which is supposedly particularly pronounced in people who retain close relationships with the natural environment' (p. 2). Those who are not in a close relationship with the natural world can more readily overlook the issue or, when faced with the enormity of concepts such as species loss or global climate change, struggle to process them. The issues can feel nebulous and overwhelming – too vast to be contended with. We look away. Empathy requires the finite and specific.

I cannot claim that practices such as attentive walking will address this issue, but I can advocate them as a way of *looking*, rather than looking away. Being present, dwelling, noticing, seeking micro details, noting tiny changes: these practises of attention and awareness enable us to connect more deeply with place. Through attention, we can also begin to appreciate time in relation to the non-human, noticing and encountering the ephemeral and the monumental: the fleeting details of small insect movement; the inevitable fall of a fragile seed head; the seasonal and cyclical; the extended presence of that which endures, in the geological bones of landscape or the steadily accumulating height and girth of trees. In his book, *Irreplaceable*, Julian Hoffman (2020, pp. 1–7) uses a detailed, vividly sensory description of a starling murmuration – taking in the season, the weather, the pier where he and others gather, the quality of light and sound on this particular evening, the people around him and their reactions to the event – to introduce the larger concept of species loss. The minutiae of a specific, place-based experience gives us access to something greater. It is from a position of localised awareness that we can attempt to approach the universal through the particular, and lay the foundations for active understanding.

Time

In developed countries, in societies that are technologically driven and geared towards capitalist, consumerist ends, our relationship to time is one of perpetual forward motion. Living in the UK, I belong to a society – and within that, an institution – that is qualified by its relationship to work. To function is to be in a state of constant busyness. This busyness is, in its turn, quantified by production: in the academy, thought is valued not for its own sake, but measured in outputs, be they learning outcomes, student assessments, teaching materials, or research publications.

In Higher Education, as in the private sector, the drive towards multimodality and the desire to integrate new technologies, while bringing benefits of (specific, technical) accessibility and efficiency, can also add to the sense of never being in one place. When divided between physical and virtual spaces, how can we be fully present? How often do we allow ourselves to be mindful, slow, present, and attentive?

In her book *Enduring Time*, Linda Baraitser (2017, p. 9), quoting Carl Cederstrom and Peter Fleming's *Dead Man Working* (2012), refers to this state of busyness as a 'flexible work force' model. The concept of flexibility might encourage us to think the model gives us more freedoms but, in reality, 'all time – work, social, leisure, family, 'quality' or unemployed time – is penetrated or 'qualified' by the logic of work' (Baraitser, 2017, p. 9). Our work time is no longer linear, in the Fordian sense of the production line; rather, all time is soaked through by work. For ease, I'm going to call this way of being 'workliving'.

Baraitser articulates the concept of 'enduring time' as a 'rupture' through this relationship with work. Periods of enduring time disrupt and destabilise workliving. Enduring time is connected to stasis, of the type we may experience during periods of isolation, illness, caring, parenting a new-born, and so on. Baraitser proposes that we 'deliberately think about staying, inertia, lack of the flow of time, lack of obvious forms of action … as a way to understand care' (p. 11). This is because care requires a different relationship to time than that experienced in the 'new chronic' (p. 13) of a society that is fuelled, and fed, by market-driven technology. For Baraitser, enduring time, which allows for care, is stuck, cyclical and repetitive; a lived experience that is 'neither eventful nor vital' but full of 'waiting, staying, delaying, enduring, persisting, repeating, maintaining, preserving and remaining' (p. 13).

Caring is what we do, as human beings, to keep the world in a suitable, sustaining state for us to flourish. By extension, care is essential to the flourishing of any academic community: in the act of study, in pedagogical practice, in pastoral and managerial roles, and in our relationship to the material environment. To varying degrees, care requires us to look after ourselves, those around us, and where we live. This implies empathy with other humans and, by a simple step of logic, a relational empathy with the non-human. In order to care – to attend to, maintain, replenish, foster – we need to embrace some enduring time. We need to slow down, rupture our fast-forward momentum, and be prepared to stay, dwell, and repeat.

When we care for something, we give it our attention. We do not look away or pass it over. In this way, attentive walking, a form of movement which is not questing, not propelled by productivity, but about noticing and attending to place, can serve a genuine purpose. It can be an act of care. Clarke and Witt echo this sentiment when defining their approach to field work, placing care alongside being and knowledge: 'Field-visiting is all about attention: caring-with attention, being-with attention, knowing-with attention' (Clarke and Witt, forthcoming 2022, p. 19).

When we are constantly *doing*, there is little sense of available free time: such moments, when they happen, can feel fallow or unsettling, outside our usual experience. We might say we feel at a momentary loss, devoid of event and purpose. Baraister identifies this as 'vague time'. She likens intervals of vague time beyond productivity to the 'vague terrains' of 'the strange in-between spatial zones in and around cities – derelict sites, empty parking lots, those bedraggled non-spaces before the city peters out' (Baraitser, 2017, p. 10). This is particularly appealing to me as a psychogeographer: what Baraitser is describing are the edgelands and unloved spaces I am often drawn to on attentive walks. Such spaces are undefined, and therefore rich in possibility. So, too, are moments of vague time: they are ideally suited to the practice of attentive walking.

Workliving and the pandemic rupture

The pandemic produced a rupture in 'chronic' time. Restrictions have come and gone, forming a sequence of connected holes, like a ladder in a stocking, through the past three years. In the UK, the first lockdown was the initial rent, a puncture. When the shock of the pandemic truly hit, many of us not involved on the frontline of medical care, or other essential forms of key work, were forced into a new relationship with time. With the busyness of our everyday suddenly cancelled, we were confined to a stuck cycle, staying put, physically and metaphorically, in a state of temporal enduring.

Businesses and institutions soon found technological fixes to the 'problem' of our de-acceleration. The result of this is a 'new normal' that takes Cederstrom and Fleming's flexible work force model to new extremes. This post-pandemic variety of workliving means that work has now inveigled itself into the nooks and crannies of what was left of our non-work lives, to the point where even periods of illness are defenceless against it. Working through Covid online at home, and working through Covid on zero hours contracts or out of fear of wage loss, are a collective step deeper into the mire of workliving. The Work Foundation at Lancaster University identifies the danger clearly:

> The pandemic has changed the way we view digital boundaries between our professional and personal lives ... An 'always on' working culture can be a major trigger and accelerator for ill health, both mentally and physically.
>
> (Taylor et al, 2022, p. 1)

That initial rupture of lockdown had the potential to teach us about our relationship with time and workliving. For some it was, briefly, an awakening, attested to by stories in the media of individuals changing their lifestyles after Covid restrictions called their attention to an unhealthy or unrewarding relationship with work. A call-in on BBC Radio 4 (You & Yours, 2021) gave voice to individuals who had chosen this moment to retire early, change jobs, downscale their hours or take pay-cuts to work in a more rewarding way, despite the prospect of rising living costs. Comments from callers discussing the pandemic's rupture through their worklives included: 'I just don't want work to be the dominant force in my life anymore' (04:13); 'I want to embed with my community' (08:47); 'It's taught me to be so much more resourceful' (12:32); 'Can't believe how naïve [I] was to work so hard for such low pay and zero benefits' (15:38); '[the pandemic] made us revaluate our working lives because we realised both the fragility and brevity of life' (27:58).

For those of us forced by lockdown into a position where we could not fully commit to workliving, our perspective on work altered, and our sense of time shifted. And when physically confined, indoors, and then on the short leash of a daily walk, so did our sense of place. We entered a stage of hyperlocalisation.

Hyperlocalisation, synecdoche and sonder

While some may have found daily walks in lockdown depressingly familiar, I found myself falling in love with the local again. Attending to my immediate environment on a daily basis, walking the same series of neighbourhood paths, I couldn't help but *notice*, and *care* about, the minutiae of those places. I began a practice of hyperlocal, attentive walking, which connected me intimately with the same locations, over and over. I attended, invested, and became increasingly entangled with specific, local pockets of place. My hyperlocal walking, initially enforced by the rupture of enduring time, and further enabled by deliberately carving out 'vague time', was an act of care both situated and cyclical. 'Distance Drift' is both a product and continuation of this determination to retain some enduring time.

Hyperlocalisation lends itself to metaphor and, in particular, to synecdoche, defined by the *Oxford English Dictionary* as 'A figure of speech in which a more inclusive term is used for a less inclusive one or vice versa, as a whole for a part or a part for a whole' (2022). Once we are attentive to the hyperlocal, we can see the specific spaces we connect with – the microhabitats of a wall we pause to study as we pass; the verges on the path of a regular walk; a particular field, spinney, allotment, waste ground or park – as a synecdoche of all such places and, by further extension, of our planet. By attending to and valuing a fragment of the world, we can better attend to and value the whole.

We may also experience a variation of 'sonder' – not only as a realisation of our own existence in relation to other humans, but to other nonhumans and places. A sonder of place. In his *Dictionary of Obscure Sorrows*, Koenig defines sonder as

> n. the realization that each random passerby is living a life as vivid and complex as your own – populated with their own ambitions, friends, routines, worries and inherited craziness – an epic story that continues invisibly around you like an anthill sprawling deep underground, with elaborate passageways to thousands of other lives that you'll never know existed, in which you might appear only once, as an extra sipping coffee in the background, as a blur of traffic passing on the highway, as a lighted window at dusk.
>
> (Koenig, 2021)

The realisation that our own uniqueness is akin to the sense of uniqueness experienced by everyone else around us – that we are all heroes of our own lives but simply bit-parts in the stories of others – leads to an understanding, as with synecdoche, that we are one fixture in a form of extended entanglement. Rich encounters with place, activated by the practice of attentive walking, remind us of this entanglement. A moment of acute, hyperlocal self-awareness can enable us to recognise the rich, entangled life of other spaces, too.

Conclusion

To attend is to care, to connect, to bring into focus and thus amplify our experience of place. As Nan Shepherd states on the closing page of *The Living Mountain*: 'Knowing another is endless. And I have discovered that man's experience of them enlarges rock, flower and bird. The thing to be known grows with the knowing' (Shepherd, 1977, p. 108).

To know, grow, and sustain ourselves, I propose that we all permit ourselves regular breaks from workliving to take stock of our immediate environ-ment. Consciously adopt periods of vague time, as a pedagogical tool and to readdress the tipped scales of your own workliving. Encourage colleagues and students to do the same; take attentive field trips around your school or college grounds. Walk if you can, slowly and with no other purpose than to embrace vague time and attend to place. Seek entanglements; take notice; care. Only by loving what is on our own doorstep can we truly care for all that lies beyond it.

References

Baraitser, L. (2017) *Enduring Time*. London: Bloomsbury Academic.

Billinghurst, H., Hind, C. and Smith, P. (eds.) (2020) *Walking Bodies: Papers, Provocations, Actions*. Axminster: Triarchy Press.

Clarke, H. and Witt, S. (2017) 'A Pedagogy of Attention: A New Signature Pedagogy for Educators', *British Educational Research Association Conference*, University of Brighton, 5–7 September.

Clarke, H. and Witt, S. (forthcoming 2022) 'Field-visiting: Paying Attention to a More-than-Human World', in Biddulph, M,. Catling, S,. Hammond L and McKendrick, J. (eds.) *Children, Education, and Geography: Rethinking Intersections*. London: Routledge.

Comtesse, H., Ertl, V., Hengst, S.M.C., Rosner, R. and Smid, G.E. (2021) 'Ecological Grief as a Response to Environmental Change: A Mental Health Risk or Functional Response?' *International Journal of Environmental Research and Public Health*, 18, 734. Available at: https://doi.org/10.3390/ijerph18020734

Hoffman, J. (2020) *Irreplaceable: The Fight to Save Our Wild Places*. London: Penguin.

Ingold, T. (2010) 'Footprints through the Weather World: Walking, Breathing, Knowing', *Journal of the Royal Anthropological Institute*, 16, pp. 121–139. Available at: www.jstor.org/stable/40606068

Knabb, K. (ed.) (1981) *Situationist International Anthology*. Reprint. Berkley: Bureau of Public Secrets, 2006.

Koenig, J. (2021) '*Sonder*' *The Dictionary of Obscure Sorrows*. Available at: www.diction aryofobscuresorrows.com/post/23536922667/sonder (Accessed 28 July 2022).

Overall, S. (2015) 'Walking against the Current: Generating Creative Responses to Place', *Journal of Writing in Creative Practice*, 8.1, pp. 11–28.

Overall, S. (2019) 'Attitudinal Walking', in Stuck, A. (ed), Glossary, *Museum of Walking*. Available at: www.museumofwalking.org.uk/glossary/?name_directory_sta rtswith=A#name_directory_position (Accessed 28 July 2022).

Overall, S. (2020) 'Pick a Drift'. Available at: www.soniaoverall.net/events/pick-a-drift/ (Accessed 28 July 2022).

Overall, S. (2021) *Walk Write (Repeat)*. Axminster: Triarchy Press.

Overall, S. (2022) #DistanceDrift 1/ [Twitter] 17 July. Available at: https://twitter.com/ SoniaOverall/status/1548593349191340033 (Accessed 27 July 2022).

Rose, M., Heddon, D., Law, M., O'Neill, M., Qualmann, C. and Wilson, H. (2022) '#WalkCreate: Understanding Walking and Creativity During COVID-19', *Walking Publics/Walking Arts*, Report Summary. Available at: https://walkcreate.gla.ac.uk/ wp-content/uploads/2022/05/WALKART-summary.pdf (Accessed 18 July 2022).

Shepherd, Nan. *The Living Mountain* (1977) Reprint. Edinburgh: Canongate, 2011.

'Synecdoche' (2022) *Oxford English Dictionary* (2022) 'Synecdoche'. Available at: www.oed.com/view/Entry/196458 (Accessed 17 November 2022).

Taylor, H., Wilkes, M. and Pakes, A. (2022) 'Digital Boundaries and Disconnection at Work: A Guide for Employers', Work Foundation, Lancaster University. Available at: www.lancaster.ac.uk/work-foundation/publications/digital-boundaries-and-disconnection-at-work-a-guide-for-employers (Accessed 28 July 2022).

You and Yours (2021) 'Call You and Yours: How Has the Pandemic Changed Your Attitude to Work?' BBC Radio 4, 19 October. Available at: www.bbc.co.uk/sounds/play/m0010pvj (Accessed 27 July 2022).

9 Entanglements

'Story telling for earthly survival'

Diane Heath and Peter Vujakovic

Introduction: an 'entangled' view of life

> It is interesting to contemplate *an entangled bank*, clothed with many plants of many kinds, with birds singing on the bushes, with various insects flitting about, and with worms crawling through the damp earth, and to reflect that these elaborately constructed forms, so different from each other, and dependent on each other in so complex a manner, have all been produced by the laws acting around us … There is grandeur in this view of life.
>
> (Darwin, from *On the Origin of Species*, 1950, pp. 414–415; emphasis added)

Darwin's poetic conclusion to his classic text assumes a human observer who is just as much part of the entanglement described as the birds, insects, and plants. That life on earth has an entangled history, and that we are a part of this has been long recognised by a wide variety of traditions and subject specialisms – for example in myth, legend and poetry, theology, evolutionary biology, philosophy, and economics. Many of these perspectives are represented in this volume.

Donna Haraway uses entanglement in a lecture, 'Story Telling for Earthly Survival' (Haraway, 2016), summing up an approach to learning to live sustainably that embraces the idea of 'natureculture' – a rejection of the binaries built up over centuries concerning art versus science, nature versus culture, and non-human against human. A similar breaking down of barriers between nature and culture, and the associated pursuit of exciting cross-disciplinary understandings is evidenced in our own approaches to learning and teaching. Haraway's idea of 'making kin' (2018, p. 103), means accepting all creatures (her term is 'critters') because we share a common 'flesh', as explored in *Staying with the Trouble* (Haraway, 2018). While recognising kinship, we must not romanticise it, from terra-forming bacteria to jungle ecosystems, we are involved with all forms of life – past, present, and becoming. We would argue that a 'good education' might draw on the principles of natureculture and 'making kin', concerned as they are with the complex entanglements between

DOI: 10.4324/9781003286516-13

humans and non-humans, permitting radical rethinking about theories of becoming, power, and agency.

Haraway (2014, p. ix) has argued that approaches tuned to 'multispecies-becoming-with' better sustain us by entering into transdisciplinary stories as ways of integrating, imbricating, and tracking threads both physical and figurative in aiding interspecies connectivity. Such approaches are one way to critically engage with and unravel the 'wicked' problems that because of inadequate information, ambiguity and changing, often contradictory demands (such as climate change or biodiversity loss), with no single, linear solution, that face the earth and all its kin today. This chapter therefore explores ideas of natureculture, kinship, and complex entanglements through two case studies, both centred on our conceptual and practical approaches to teaching entanglement in higher education and in the community.

Prior to higher education many students have mainly worked in silos defined (in the English context) by the General Certificate of Secondary Education and A Level curriculum (and equivalents). There the tendency is to separate out the sciences from the humanities and arts. Natureculture approaches seek to reinstate a more holistic understanding, for example, the first case study examines the value of a phytobiographic engagement with trees, a project in long-life learning – thinking in 'tree-time' noting that student's critique accepted conceptualisations of trees as static and largely benign 'critters' and instead regard them as much more complex and 'gnarly' agents. The second turns to native oysters (*Ostrea edulis*) as figures of resilience and kindred mutual support, in a kinaesthetic approach to teaching about a particular entanglement – 'holobiomes'.[1] These approaches located within higher education seek to disrupt the silo learnings often developed prior to entering tertiary education.

Gnarly agents: entanglements with trees

> Yet portion of that unknown plain
>   Will Hodge for ever be;
> His homely Northern breast and brain
>   Grow up a Southern tree,
> And strange-eyed constellations reign
>   His stars eternally.

The third stanza of Thomas Hardy's Boer War poem, *Drummer Hodge*, is a grim reminder of our livelong entanglement with trees. A drummer boy is buried 'uncoffined' on the South African veldt and a tree arises from his decay. For much of humanity's existence trees have been critical to our daily lives, providing food and shelter, then heat, tools, and finally our casket. Today their role remains critical in terms of sustainable ecosystem services, climate change mitigation, and enhanced biodiversity. These objective criteria are enhanced by the potential for students to engage with trees as individual organisms with distinctive life forms and life histories. Trees have been used by the author (PV) to elucidate key concepts in sustainability across multiple modules, through class and field learning experiences.

This section examines entanglements with trees as non-human agents via a 'phytobiographic'[2] approach (Vujakovic, 2013, 2019a and 2019b). It acknowledges trees as long-lived organisms that may have come into existence before we were born and will thrive long after we are gone. These are organisms that we can engage with individually and come to understand their life-histories. Their relationship with humans as related evolved organisms is explicitly discussed in Richard Dawkins's (2004) *The Ancestor's Tale: A Pilgrimage to the Dawn of Life*. His narrative is a trenchant argument for a view of life as universal kinship with a common 'concestor' – the grand ancestor of all life – on the way to which we 'rendezvous' with the plants, which Dawkins names 'the true lords of life' (p. 518). Dawkins's narrative is also an incisive reminder that we should not see the story of life on earth as leading to 'us', but that we will inevitably contemplate our relationship with non-human nature from our own perspective.

It should be noted that trees are not a distinct category like 'cats' or 'antelopes', but a strategy adopted by very wide range of plant groups; a form of convergent evolution where different plants have adopted the same or similar solutions to life's problems – in this case harvesting light by extending your photosynthetic apparatus skyward on a tall trunk.

The longevity of trees affects the way we (as individuals and society) might envisage entanglements with them, offering interesting lessons in 'long-life learning' – thinking in 'tree-time' (Vujakovic, 2013). Entanglements with trees involve long time horizons. Owners of estates knew they would never see the culmination of their laying out of avenues of trees or woodlands – they were working for future generations (intergeneration as well as inter-species entanglements). Where human entanglement is with mature trees it is complex in temporal terms: from daily rhythms of work and rest to seasonal factors, and to yearly and decadal variations in fruit and wood production, disease, and storm damage.

Tim Ingold (1993) discusses this relationship with trees in reference to Pieter Bruegel the Elder's painting, The *Harvesters* (1565). Ingold uses the painting to unfold his concept of the 'taskscape' – landscape as the product of habitation and use. He notes regarding a pear tree central to Bruegel's scene:

> Rising from the spot where people are gathering for their repast is an old and gnarled pear-tree, which provides them with both shade from the sun, a back-rest and a prop for utensils. Being the month of August, the tree is in full leaf, and fruit is ripening on the branches … [its] history consists in the unfolding of its relations with manifold components of its environment, including the people who have nurtured it, tilled the soil around it, pruned its branches, picked its fruit, and – as at present – use it as something to lean

against. ... [I]t presides immobile over the passage of human generations ... and the regular round of human agricultural activities.

(pp. 167–168)

In Ingold's discussion the tree is a benign, passive, and immobile critter. There is no reference to its agency; rather, it is something that is done to – pruned, harvested, and lent against. This misses the fact that trees represent a curiously alien form of life-agency so unlike humans. Tied to one place once they have germinated (unless moved by humans), they can shed limbs with no serious impact to health, communicate chemically, often form 'clonal colonies', live for centuries, and many are immense in height and bulk. There is no denying, however, their importance as non-human actors with agency, despite their apparent immobility, and the fact that they are highly competitive organisms.

Decades of teaching about trees has identified them as agents in the landscape, aptly characterized as 'gnarly critters'; as organisms that are not always benign as supposed but aggressive agents that compete with other organisms for space, access to light, water and nutrients and the extension their gene pool (see Figure 9.1).

'Gnarly' (synonym 'gnarled') is the perfect word for trees as active agents. The first, and formal meaning is straightforward; its common usage is to mean twisted and knobby, used to describe tree trunks, roots, and branches, as well

Figure 9.1 Veteran trees, Staverton Thicks, Suffolk.

as the human body, particularly hands and feet. Its first use is attributed to Shakespeare in *Measure for Measure* (Act II, Scene II) with reference to a 'gnarled oak' (possibly a variant on Middle English '*knurled*' for knotty wood – OED). Thomas Hardy, a master of arboreal description, makes the human–tree connection wonderfully in his description of the scene that meets Gabriel Oak's eyes as he steps into Warren's Malthouse in Chapter VIII of *Far from the Madding Crowd*:

> A curved settle of unplaned oak stretched along one side, and in a remote corner was a small bed and bedstead, the owner and frequent occupier of which was the maltster.
>
> This aged man was now sitting opposite the fire, his frosty white hair and beard overgrowing his gnarled figure like the grey moss and lichen upon a leafless apple-tree.
>
> Chapter VIII, p. 46 (1872, pagination from
> Penguin Classics edition, 2000)

The other common use of 'gnarly' (slang) means difficult, as in a 'knotty' or 'thorny' problem' or more generally 'bad' or 'unpleasant'. Finally, its use has been turned on its head (in the way 'bad' can mean good) and has come to mean 'excellent'. These last two counter-meanings draw attention to the complexity of trees as active agents – they are both a source of the eco-space and resources that sustain numerous other organisms as well as humans, they are also aesthetically pleasing, but their role as competitive organisms which can crush, out-shade and even poison their competitors fits the negative connotation, too.

The rest of this section explores this ambiguity and lessons for sustainability through case studies based on teaching and research at Canterbury Christ Church University with a focus on trees as gnarly agents that can confound humans as well as oblige them. The case studies are also treated as examples of tree/human entanglements within 'novel ecosystems', a concept that accepts that most ecosystems have been transformed 'into new, non-historical configurations owing to a variety of local and global changes … *hybrid systems* retaining some original characteristics as well as novel elements' (Hobbs, et al., 2009, p. 599; emphasis added).

Space invaders – the Walmer holm oaks

Walmer beach is a stretch of shingle running north–south along the coast of Kent (UK). The shingle is bounded on one side by the English Channel and on the other by an esplanade, road, and housing. The beach appears to be a thin line of wilderness – a liminal space between town and sea. What is striking about the beach are the huge dome-shaped growths hugging the shingle, often several metres high and up to ten metres across. These are holm oaks (*Quercus ilex*), shaped into aerodynamic growth-forms by their location on the edge of the sea, at the mercy of harsh winds. These distinctive organisms have the advantage of their enormity; they have been the subject of both physical field-work, and subsequently of teaching using remote sensing (air photographic evidence; including longitudinal data sets from the 1940s onward) and other 'virtual' evidence (Google Street View) during COVID lock-down. While less than perfect, virtual engagement could still provide opportunities for entangled learning as these organisms revealed their narratives.

Usually, an elegant tree with a distinct trunk, the Walmer oaks have adopted a 'hunker-down' life-form through which the upper beach resembles a landing ground for arborescent flying-saucers! A native of the Mediterranean, widely naturalised throughout Britain, holm oaks have adapted to a range of habitats where their ability to withstand harsh, dry conditions give them an advantage. At Walmer they have colonised a space that is inhospitable to most native trees, creating a novel ecosystem full of contradictions in tree–human terms. The holm oak is one of several invasive plant species (mostly ground flora) able to cope with extremes of exposure, and low nutrient and moisture availability on the shingle.

Air imagery from the 1940s to the present provides evidence of this invasion, its impact, and the phytobiography of individual trees. The initial incursion was some 600 metres to the south of Walmer Castle. The castle grounds contain mature holm oaks (original planted in the 1860s), the putative source of the beach population. An area of scrub behind the esplanade appears to be the route by which the oaks reached the beach. Several holm oaks are well established there, probably dispersed by birds (zoochory) from the castle grounds. *Chory*, the term for seed dispersal, was originally coined by Dammer in the 1800s and is derived from *chorein* (Greek) – 'to wander' (Van der Pijl, 1982; p. 7). The conundrum is how these oaks crossed the esplanade to reach the beach. There would be little incentive for a bird or squirrel to carry acorns across open shingle toward the sea. One intriguing thought is that the dispersal mechanism might be play! Children notoriously collect, kick, and throw 'nuts' including acorns and 'conkers' (horse chestnut – *Aesculus hippocastanum*). We might designate this dispersal method 'playful-wandering' or *paizeinchory*; adding *paizein* from the ancient Greek 'to frolic' (Banchich, 2017). Once trees had been established in this manner, bird dispersal could commence as the trees matured and became a viable food resource.

One very large holm oak, just to the landward side of the esplanade, appears to be at the epicentre of dispersal to the beach (see Figure 9.2). Air photography shows this oak to have been well established in the 1940s, and the invasion of the beach then commences north and south, until there are now nearly eighty well-developed oaks on the shingle. These are dominating organisms imposing themselves on the ecosystem – forming a hybrid shingle–savanna habitat. Their crowns are dense and evergreen and reach to the ground around most of their perimeter. This suppresses (by dense shade) any vegetation beneath, while dead oak leaves form a mulch that adds to the effect. The interiors are an open

Figure 9.2 Holm oak on Walmer Beach, Kent.

structure of radiating branches, 'gnarly' dens and climbing structures for children – further playful entanglement.

What the air images also reveal is another sense in which these oaks could be characterised as active agents and space invaders: around each is a very distinctive 'halo' where the shingle is devoid of any other vegetation (Finn and Vujakovic, 2020). Extending beyond the crowns the roots take all the available moisture and nutrients from this impoverished habitat. Even the highly adapted valerian (*Valeriana officinalis*), which grows on the shingle, is unable to grow within the oaks' sphere of influence. The halo effect has 'gnarly' consequences for the native flora of the shingle, especially as the oaks spread further along the coast. This provides an object lesson in understanding trees as aggressive non-human agents.

Differences in halo effect seem to be partly related to the impact of the urbanised sections of the beach. A problem identified by the White Cliffs Countryside Partnership (2010) is dumping of garden waste onto the beach. This encourages invasion by other exotic species, and nutrient enrichment of the shingle habitat to the detriment of native species. The air imagery shows this issue clearly, with lush ground cover on the shingle immediately to the front of homes, compared with much more exposed shingle to the north, where no housing exists. Large holm oaks close to housing seem to have much less impact due to the nutrient enrichment and the fact that the species surrounding them, including grasses, are adapted to making use of this. The other 'gnarly' issue is some residents' dislike of the holm oaks, which they feel obstruct their view of the sea, and this leads to calls for their destruction (White Cliffs, 2010). The genetic plasticity of trees is clearly evinced by these oaks, crouching in their modified 'cushion' form. Good for them, but not for the native species of this important and rare habitat type and an ambiguous relationship with humans as 'kin'.

The living and the dead – trees in burial grounds

Burial grounds may seem odd places to celebrate entanglement with the living world, but their value as urban green space is well recognised, especially with the growth of the sustainability agenda. As David Goode (2014) notes: 'Ask any urban wildlife trust for a list of its best nature areas and it will probably include one or two cemeteries'. (p. 84). Central to most of these spaces is their habitat form as 'urban savannas' – archetypal novel ecosystems – mosaics of open grassland and scattered parkland trees. Cemeteries are spaces of both positive and negative gnarl entanglement, with veteran trees providing biodiversity and aesthetic value, but also acting as invasive and potentially destructive agents. Cemeteries are amongst the most important accessible green spaces in most towns and cities, a valuable resource for teaching and learning for sustainability and for the understanding of 'green heritage' (Vujakovic, 2018). Green heritage is an approach that values nature as well as built heritage and seeks to overcome inherent contradictions where the two elements may be in conflict (see below).

Cemeteries and graveyards often contain some of the most important tree collections in urban areas (Vujakovic, 2019a and 2019c). Particularly important are the great Victorian and Edwardian cemeteries of the mid-nineteenth and early twentieth centuries. Originally businesses (joint stock companies), the owners created beautifully planted landscapes to attract custom (Brooks, 1989). These cemeteries were to cope with the population boom in industrial cities and designed to become, when full, botanic gardens or arboreta – as places of memorial, recreation, and education. One of the most influential designers of the Victorian period (see Curl, 1983), John C. Loudon (1843), was explicit that cemeteries should serve as 'breathing spaces' and 'future public gardens'. Loudon also wished to create a sense of solemnity, so for both aesthetic and practical reasons he advocated evergreens (pines, firs, yews, and similar species – many exotic) over native deciduous trees. These evergreen species were easy to shape or grew naturally in conical form; they did not create problems of leaf clearance; and they retained their shape in winter. Older churchyards and many cemeteries (via self-seeding from nearby sources) do contain large veteran natives. Veteran trees often exhibit a remarkable range of gnarly-microhabitats that attend the aging process, including rot hollows, bark cavities, dead wood, water pools, and sap runs. These trees provide excellent

Figure 9.3 Irish yew trees, village churchyard, Kent.

teaching resources and opportunities to train students in identification of microhabitats and their significance for biodiversity (see Figure 9.3).

Some of the ecological interest of burial grounds, especially large urban cemeteries, is the product not of deliberate planting, but of neglect, significantly from the late eighteenth century onward. While lucrative investments in their early years, privately owned cemeteries experienced declining management as costs rose. Other gnarly problems contributed; Goode (2014) notes that 'the death knell for most of the cemeteries was World War I when gardeners were no longer available, and nature ran riot' (p. 91). Areas where interments were still taking place were given priority, while other parts succumbed to scrub and trees incursion. This has, however, proven a double-edged sword, with some increase in biodiversity, but also damage to monuments associated with the rampant growth of secondary woodland species, primarily the easily dispersed, winged-seeded ash (*Fraxinus excelsior*) and sycamore (*Acer pseudoplatanus*). These and other species, like bird-dispersed holly (*Ilex aquifolium*) and yew (*Taxus baccata*), have taken advantage of tombs and headstones to provide protection from grass-mowers and strimmers, chief tools of management. Thus, monuments have sown the seeds of their own destruction by providing protection for seedlings – literally turned to piles of rubble by the agency of trees.

Many cemeteries now have 'friends' groups that try to balance the thorny problem of heritage preservation versus nature conservation; they often work alongside local authorities, who have overall managerial control. A fascinating study of Arnos Vale Cemetery (Bristol) examined the complex, gnarly-entanglements between humans and trees (Cloke and Jones, 2004). For developers who wished to clear the site, secondary woodland was evidence of neglect and abandonment, while for locals it represented valued green space. Cloke and Jones explicitly regard the invasive trees as 'non-human agents[,] … as active participants in (re)making place'. They also note that the trees have played a significant part in the way local people have resisted development, undertaking a tree survey in which trees were mapped in relation to pathways, which contributed to a management plan. This in turn was the basis for community 'Tree Gazing Walks' that invigorated wider interest in the site and its ecology (Jones, 2007). The necropolis can offer exceptional educational insights into the gnarly entanglement between trees and humans.

Lithic agents: entanglements with oysters

The oysters in their rocky beds along the freshets and shallows of the Thames estuary have been cultivated since mesolithic times. In the Middle Ages, the shellfish provided both a source of income, for they were sold in local and London markets, and staple food for Kent maritime communities. Thames oyster beds were decimated at the start of the twentieth century by overfishing, pollution, and the introduction of new predators, and have suffered more recently, as the seas warm and population pressures on coastal habitats increase; now less than 5 per cent of the original habitat remains since the mid-nineteenth century (Native Oyster Network website, 2022). This is a terrible loss. Yet in the liminality of their littoral locations and their literary and symbolic associations the natureculture entanglements of the native oyster (*Ostrea edulis*) still enrich our lives and, like Peter's gnarly trees, provide enriching educational experiences. Oysters are coastal, communal, and turn out to be considerably different to how we generally perceive them.

As an example of differing perceptions (some of which linger in our cultural memory banks still), in the Middle Ages it was believed that the oyster, more stone than animal, made her pearl not from detritus but from heavenly light. She (then all oysters were believed to be female) would swim to the surface of the sea and open her shell at daybreak to receive a drop of dew suffused with sunlight, moonlight, and starlight that formed the core of her pearl. Pearls were a symbol of purity, for as gifts of pure teardrops from heaven, they represented the Christ child given to the Blessed Virgin Mary and, by analogy, all children became heaven-sent. In the anonymous medieval poem, *Pearl*, the symbolism extends still further – the pearl is a child, yet also a young woman, and also a spotless gemstone that is both the human soul and heavenly Jerusalem, in a poem that is about the loss of a beloved child.

> Perle, plesaunte to prynces paye
> To clanly clos in golde so clere,
> Oute of Oryent, I hardyly saye,
> Ne proved I never her precios pere.
> So rounde, so reken in uche araye,
> So smal, so smothe her sydes were,
> Queresoever I jugged gemmes gaye
> I sette hyr sengeley in synglure.
> Allas, I leste hyr in on erbere;
> Thurgh gresse to grounde hit fro me yot.
> I dewyne, fordolked of luf-daungere
> Of that pryvy perle withouten spot.
>   (Anon., *Pearl*, 14th century, lines 1–12)

The pearl in its luminescent roundness was considered as the orb in Christ's hand, a globe representing the world, and a cure for poison. In sympathy, the oyster was the rock representing the medieval Church, a lithic St Peter, guarding the pearly Jerusalem. The outside oyster shell is grey, often barnacled and bumpy, but the inside is shiny, sleek, and smooth – aptly-named mother of pearl. Those links to medieval Marian perfection of motherhood carried on into the seventeenth century, when Marcus Gheerhaerts II painted heavily pregnant women bedecked in pearls (Gheerhaerts II, c. 1590s, oil). Despite the evocative nature of these traditional cultural readings of the medieval pearl, the materiality of the oyster lifecycle disrupts such understandings. In this section, I (DH) demonstrate how this disruption can bring us closer to an exploration of the liminal spaces between sustainability and learning.

The lifecycle of the native oyster (*Ostrea edulis*) is very different from these medieval (mis)conceptions. Oysters are not only female, but male too; that is hardly a surprise. However, at the beginning of the twentieth century it was discovered that native oysters are alternating hermaphrodites. Born male and born alive (around a fortnight after conception), as adults they in turn become female and back to male depending on factors such as seawater temperature. Female oysters release millions of microscopic live larvae, 95 per cent of which will die. The survivors float in the shallows until they grow a foot-like appendage that allows them to stick to rocks or other oyster shells, forming bed-like reefs, where they remain for up to twenty years. It is a precarious lifestyle; 95 per cent of larvae die, and the species' population can be adversely affected by an imbalance of gender ratios as well as pollution and ocean warming (Native Oyster Network).

By building on the shells of their ancestors and siblings the oysters create rock-like holobiomes (like coral reefs), providing rocky shelters in the salty shallows for schools of young fish, sponges, and squirts, crustaceans, and other shellfish. This inclusive habitat, at once liminal (sited between sea and shore) yet linked to mainstream aquaculture, turns into a learning space for us today. Not only do their reefs host multispecies diversity, they also clean the seawater they filter by increasing its clarity and removing pollution such as nitrates. As an example of how oyster holobiomes have become 'learning spaces', the Native Oyster Network (NON) has produced posters to advertise the oysters' 'services' – from decreased turbidity to better water quality to enhanced biodiversity and cultural impacts on maritime communities, and NON even had a photographic competition for the most beautiful oyster. Marine biologists and researchers are therefore eager both to protect the surviving natural locations and to try and recreate such habitats in new places using artificial 'bio-huts' in harbours and under piers.

Of course, it is not just for humanity, but for multispecies dependency that we need to act now: it is urgent for every what and every who that lives on the planet. One way to start would be to stop seeing oysters as merely seafood but instead as vital parts of globally beneficial biodiverse assemblages. Taking my inspiration from Haraway, I explored 'making kin' in two ways.

Community entanglements

The first form of engagement comprised children's activities sessions at a heritage hub exhibition in the Town Hall of the medieval oyster fishing port of Faversham. The heritage hub setting was part of a Faversham Town Council Funded project with Canterbury Christ Church University's Centre for Kent History and Heritage, to celebrate the medieval history of this Kent port (Richardson, 2017) and its unique collection of medieval charters and archival collections. At 'Open Saturdays' the Heritage Hub included a 'Young Medievalists' Corner', where I hosted with colleagues free Key Stage 2 educational activities for local children (typically 7 to 11 year olds) to 'learn through play'. I developed an activity that connected the history and blue and green heritage of Faversham's medieval oysters (then a vital part of the town's economy) to current understandings of biodiversity and sustainability. Children were told the story of how medieval people believed pearls were made from light, that oyster beds have become severely depleted, and that we should value those left. Recycled oyster shells (from catering suppliers) were reused by drilling a hole to take a small slipring. Children were encouraged to make their own 'pearls of wisdom' in response to the heritage exhibition. Each child's messages were then stuck on their chosen oyster shell by a glued pearl for them to display at home as a keepsake using a ribbon attached to the slipring. Messages included 'Medieval Faversham', 'I love adventure', and some children wrote their names, making themselves the medieval heavenly pearl gift. Several children said their keepsakes would be given as gifts to their parents, while others left wearing their makings as necklets; both uses indicate the children had learnt about their town's cultural heritage and also about valuing their local (sea) blue heritage.

Researcher entanglements

A second group took part in a 'hands-on' kinaesthetic workshop at the Gender and Medieval Studies conference at Swansea University in January 2020. Following a paper on medieval oysters and the representation of the natural world (Heath, 2023 forthcoming), delegates fashioned oyster shells with 'pearl of wisdom' messages for themselves and friends. These keepsakes became mementoes of the pre-Covid Swansea seaside, remembrances of academic friendships, and several delegates have their oystershells over three years later, a significant period of time, especially since it included the pandemic, when lockdowns prevented in-person meeting and conferences. The making activities in both settings were apt tasks as they emphasised creativity, cooperation, and support, mirroring the oysters' mutual holobiome support structure.

Haraway's slogan of 'Making kin not babies' has opened up medieval perceptions and preconceptions of the triple themes of sanctity of motherhood, fecundity, and the purity of the pearl to gender fluid understandings of togetherness. For example, gender-alternating oysters cannot know their father, nor recall their mother but have millions of siblings. Learning their lifecycle provides insights by paying 'serious attention to multispecies flourishing', and by comparison, highlights 'issues of race and class and region for human beings, as means of making kin, not babies' (Franklin, 2017, p. 61). As alternating hermaphrodites, oysters' gender fluidity means turning from symbolising the womb to becoming figures of mutual support in biodiverse reefs. Learning their lifecycle provides insights into issues such as climate emergency and social justice. The deceptive simplicity of making recycled oyster shell keepsakes allows both children and adults to learn of the fragility of the world and brings the plight of damaged and depleted oyster beds closer to our thoughts and feelings.

Conclusion – entangled learning

Our experience of connecting with non-human actors through our learning, our teaching and community engagement has reinforced the need to see this relationship as complex, entangled, and gnarly. Teaching about sustainability requires discussion of wicked problems and acceptance that entanglement with our kin is both a reminder of our common origins, the complexity of environmental systems, and that living creatures are 'always in the business of surviving' (Dawkins, 2004, p. 5) and may act against our interests.

Survival does not always involve positive or benign interactions between kin. Entangled and long-life learning requires reference to the past (including an understanding of processes in 'deep-time'), present and future, to science and culture, and a sense of humility. While valuing and celebrating our kinship with other critters, and working toward a sustainable world, we must maintain a sense of perspective and not fall into the trap of a romanticized view of our entanglements.

Our joint chapter also recognises our shared and wider interest in 'blue' and 'green' heritage, an approach that values nature as well as built heritage. Green heritage involves nurturing or restoring the living element within heritage spaces and 'blue heritage' respects and sustains our historic salt and fresh water areas. Both 'blue' and 'green' heritage also involve interpretation of these sites via reference to the living world, both in the past and the present. This merging of past present and future and the emphasis on biodiversity resonate with Haraway's natureculture.

Notes

1 Holobiome – an assembly including a host (for example coral and oyster beds) and other species living in or around it that produce a distinct ecological entity.
2 The phytobiography approach acknowledges the fundamental physiological and ecological drives that determine plant growth while also recognizes that each organism has a 'life-story', a narrative that facilitates deeper understanding of our relationship to the living world.

References

Anon., *Pearl*, c. 1375–1400, ed. Stanbury, S. (2001) Kalamazoo, MI: MIP, TEAMS.

Banchich, T. (2017) 'A Gag at the Bottom of a Bowl? Perceptions of Playfulness in Archaic and Classical Greece'. *American Journal of Play*, 9(3), pp. 323–340.

Brooks, C. (1989) *Mortal Remains: The History and Present State of the Victorian and Edwardian Cemetery*. Exeter: Wheaton, in assoc. with The Victorian Society.

Cloke, P. and Jones, O. (2004) 'Turning in the graveyard: trees and the hybrid geographies of dwelling, monitoring and resistance in a Bristol cemetery'. *Cultural Geographies*, 11, pp. 313–341.

Curl, J.S. (1983) 'John Claudius Loudon and the Garden Cemetery Movement'. *Garden History*, 11(2), pp. 133–156.

Dawkins, R. (2004) *The Ancestor's Tale: A Pilgrimage to the Dawn of Life*. London: Phoenix.

Darwin, C. (1950) *On the Origin of Species by Means of Natural Selection* (reprint of the first edition published in 1859). London: Watts & Co.

Finn, L. and Vujakovic, P. (2020) 'Space Invaders: the holm oaks of Walmer Beach'. *City Trees*, 57, pp. 32–37.

Franklin, S. (2017) 'Staying with the Manifesto: An Interview with Donna Haraway'. *Theory, Culture & Society*, 34 (4), 49–63.

Gheerhaerts II, M. (c. 1590s) *Unknown Lady* (oil). London: Tate. www.tate.org.uk/art/artworks/gheeraerts-portrait-of-an-unknown-lady-t07699.

Goode, D. (2014) *Nature in Towns and Cities*. London: HarperCollins (New Naturalist).

Haraway, D.J. (2014) 'Preface' in: Nagai K., Rooney, C., Landry, D., Mattfeld, M., Sleigh, C., and Jones, K. *Cosmopolitan Animals*. Houndmills: Palgrave Macmillan, pp. vii–xii.

Haraway, D.J. (2016) *Story Telling for Earthly Survival* (film dir. Fabrizio Terranova)

Haraway, D.J. (2018) *Staying with the Trouble*. Durham, NC: Duke University Press.

Hardy, T. (2000) *Far from the Madding Crowd*, London: Penguin Classics (First publ. 1874).

Heath, D. (2023) 'Recreating the 'Natural World': The Medieval Oyster and her Pearl', in *Gender, Science and the 'Natural' World* (eds.) T. Tyers and P. Skinner. Cardiff: University of Wales Press.

Hobbs, R.J., Higgs, E., and Harris, J.A. (2009) 'Novel ecosystems: implications for conservation and restoration', *Trends in Ecology & Evolution*, 24(11), pp. 599–605.

Ingold, T. (1993) The Temporality of the Landscape', *World Archaeology*, 25(2), pp. 152–174.

Jones, O. (2007) Arnos Vale Cemetery and the Lively Materialities of Trees in Place, *Garden History*, 35, Supplement: Cultural and Historical Geographies of the Arboretum, pp. 149–171.

Loudon, J.C. (1843) *On the Laying Out, Planting, and Managing of Cemeteries, and on the Improvement of Churchyards*. London: Longman, Brown, Green and Longmans.

Native Oyster Network (2022) website https://nativeoysternetwork.org/resources/ photo credit Dr Paul Naylor (accessed 29 July 2022).

Richardson, A. (2017) ''What Came Before': The Kingdom of Kent to AD 800', in Sweetinburgh, S. (ed.) (2017) *Early Medieval Kent, 800–1220*. Woodbridge: Boydell & Brewer.

Van der Pijl, L. (1982) *Principles of Dispersal in Higher Plants*, 3rd edn. Berlin: Springer-Verlag.

Vujakovic, P. (2013) 'Phytobiography: an approach to "tree-time" & "long-life learning"'. *Arboricultural Journal*, 35(2), pp. 134–146.

Vujakovic, P. (2018) 'Necropolis mapped: geography, education and the cartography of cemeteries'. *Bulletin of the Society of Cartographers*, 53, pp. 25–34.

Vujakovic, P. (2019a) 'Engagement with trees as long-life learning for sustainability', in Fihlo, W. (ed.) *Encyclopaedia of the UN Sustainable Development Goals: Life on Land*. New York: Springer Reference.

Vujakovic, P. (2019b) 'Knockout blow? Phytobiography revisited', *City Trees*, 56(4), pp. 25–31.

Vujakovic, P. (2019c) 'A Matter of Life and Death: Trees in city cemeteries and graveyards', *City Trees*, 56(5), pp. 16–21.

Werner, M. and B. Zimmermann. (2006) 'Beyond Comparison: *Histoire Croisée* and the Challenge of Reflexivity'. *History and Theory*, 45(1), pp. 30–50.

White Cliffs Countryside Partnership (2010) *Kingsdown and Walmer Beach, Draft Management Plan 2010–2014*. Available at: https://kingsdownconservationgroup. files.wordpress.com/2010/12/kingsdown-and-walmer-beach-management-plan.pdf (Accessed: 5 February 2023).

10 Understanding sustainability pedagogy in practice

John-Paul Riordan

Introduction

Global crises due to human activity, such as ecological imbalance, the climate crisis, and widening economic and social disparity means many endure, or will face, hardship (Pörtner et al., 2022). As a teacher educator and pedagogy researcher in higher education, I am challenged as to how to help trainee teachers educate children regarding complex sustainability issues. My starting point to address this challenge is to focus on understanding what is actually happening in classroom contexts. This chapter seeks to explore the ways teachers and learners respond to the ethical questions that arise during school lessons in relation to sustainability. Two related questions are explored: what is sustainability pedagogy, and how might classroom pedagogy research contribute to Initial Teacher Education regarding sustainability? To answer both questions and untangle complicated pedagogical interactions I employ a theory referred to as the Pedagogy Analysis Framework (PAF) (Riordan et al., 2021). The chapter begins by briefly reviewing literature and situates sustainability pedagogy as a type of ethical pedagogy. I then analyse a real example of sustainability to illustrate the role of the PAF in exploring classroom-based pedagogical interactions. Using excerpts from a UK primary school Religious Education lesson, with pupils aged 8–9 years old, on sustainability in a topic about 'Creation Stories', I show how the PAF can develop understanding of sustainability pedagogy in practice.

Understanding sustainability pedagogy – the Pedagogy Analysis Framework (PAF)

Pedagogy is itself a contested term. It can focus on what teachers do in classrooms, for example, Alexander notes that

> pedagogy is the act of teaching together with its attendant discourse of educational theories, values, evidence, and justifications. It is what one needs to

DOI: 10.4324/9781003286516-14

know, and the skills one needs to command, in order to make and justify the many different kinds of decision of which teaching is constituted.

(Alexander, 2008, p. 47)

In contrast, others understand pedagogy to involve the interactions between pupils and teacher: '[Pedagogy involves] those factors affecting the processes of teaching and learning and the inter-relationships between them (Hallam and Ireson, 1999, p. 78, quoted in Black and William, 2018, p. 555)'.

Both definitions are helpful, but I would also include interactions between any participant (pupil or teacher) and the material world (Hardman, Riordan and Hetherington, 2022). For example, how a pupil interacts with equipment in the classroom obviously influences learning, to a greater or lesser degree, and the sort of support learners may need. Hence, my understanding of pedagogy encompasses the interactions of participants (be they teacher, pupil, or researcher) with each other, interactions of each participant with themselves – for example with their own thoughts, feelings, motivations, and actions – and interactions between any participant and the physical objects in the room, be they real or imaginary (see Hardman, Riordan and Hetherington, 2022). This complexity is intensified when considering the relationship between pedagogy and sustainability. The etymology of 'sustain' can be 'to hold from below' but, according to White (2008), can also involve 'sustenance' or 'enduring adversity'. Consequently, a diverse range of (subtly different) descriptors are deployed including sustainable pedagogy, pedagogy for sustainability, education for sustainability, and education for sustainable development (e.g., see Salas-Zapata, Ríos-Osorio & Cardona-Arias, 2018). The present chapter uses the term *sustainability pedagogy* to mean any interaction between people, or between people and things, involving avoiding depletion of natural resources so as to maintain ecological balance (drawing on the Organisation for Economic Co-operation and Development definition). Sustainability can appear in formal curricula (for example the secondary science content of the UK National Curriculum discusses climate change), can emerge informally during lessons (DfE, 2014), and has recently been highlighted in UK education policy (DfE, 2022). Hence, this definition of sustainability pedagogy argues that at any instant of a lesson multiple elements of the pedagogy can be concerned with sustainability. Kopnina and Meijers (2014) argue classroom ethnographic research regarding Education for Sustainable Development is needed and currently rare.

I think anything as complicated as sustainability pedagogy needs to be analysed from multiple perspectives. The extended Pedagogy Analysis Framework is an emergent Grounded Theory (Riordan, 2022; Riordan, Hardman and Cumbers, 2021; Riordan et al., 2021) and is one way to understand and explain the elements of pedagogy and how they relate. It involves detailed analysis of classroom pedagogy using video data from real-life lessons that took place in school contexts. This interpretivist approach complements those bringing a critical theory lens to sustainability pedagogy (discussed elsewhere in this book). In the following section, I analyse a short transcript of video data from a lesson on 'Creation Stories' from one UK primary school to illustrate the potential value of the PAF.

Analysing a real example of sustainability pedagogy

A short transcript from a UK primary school Religious Education lesson, with pupils aged 8–9 years old, about sustainability in the topic 'Creation Stories', provides the context for the analysis. The lesson was video recorded from three angles. Then the class teacher watched the whole lesson back whilst being videoed 'thinking aloud' (Taylor and Dionne, 2000). Next, the pupils watched clips of the lesson and were also videoed as they talked together. The complicated nature of sustainability pedagogy requires analysis from multiple perspectives and the Pedagogy Analysis Framework, outlined next, helps with detailed analysis of the micro aspects of classroom pedagogy.

The following short interaction from a UK primary school lesson between a teacher (Ms Ivy), and two pupils, both aged 9 (Oliver and George, pseudonyms), features the teacher's question about sustainability, three solutions from Oliver to this sustainability problem (pollution of the natural environment), the teacher's response, and George's response. The aim is to illustrate sustainability pedagogy with a real example and discuss how classroom research like this can contribute to Initial Teacher Education regarding sustainability. Initial Teacher Education students, Early Career Teachers (ECT), and school coaches may sometimes find it helpful, particularly when using video for professional development, to use the PAF like this to untangle complicated classroom pedagogy. This lesson is included as it contains many sustainability-related interactions. To provide context, the pupils have just watched, in the classroom, Michael Jackson's 'Earth Song' music video which addresses (controversially) many sustainability issues, including pollution.

Ms Ivy:	How do you think you could make a difference to our world? Oliver?
Oliver:	In making like … So [starts to hold up his thumb to indicate a list], pollution-sucking pigeons, plant … plant … plant … planting vines, and plastic eating tur… robot turtles. And then they could get rid of all the stuff.
Ms Ivy:	So … OK … so you think that maybe, what if you grew up to be a scientist and you created … [Oliver interrupts].
Oliver:	I am going to be.
Ms Ivy:	Are you? Well … [Oliver interrupts again].
Oliver:	I am going to be an inventor who creates a whole chocolate factory.
Ms Ivy:	Fantastic. Sounds amazing to me Oliver. Do you know Oliver, if you can use science to help, like you said about those robots that can clear the plastic from the sea. What could we do though, just us? It is amazing that Oliver wants to be an inventor and he wants

to create this robot, but what could we do with the plastic to start with? What could we do George?

George: Pick all the litter up and put it in the bin. Also, on Christmas Day, in a war, people, just for one hour, German and English people, stopped fighting, put their guns down and people from Germany and England came out of their trenches and they just all played football.

In response to Ms Ivy asking the whole class about what they could do to make a difference by being more sustainable; Oliver offers three imaginative solutions to the pedagogical problem set by the teacher of what they personally could do about pollution (i.e., breed pigeons to remove air pollution presumably by inhaling it, deploy plastic-eating robot turtles, and use plants), and notes these sustainability solutions would, 'get rid of all the stuff'. Ms Ivy begins summarising Oliver's points by positioning his solutions in the future, when he will have become a professional scientist. Oliver interrupts to tell Ms Ivy that he *will* be a scientist, which Ms Ivy acknowledges with a rhetorical question. Oliver then clarifies that he will be an inventor and introduces his chocolate factory plan. Ms Ivy acknowledges these plans with praise three times, points out this is using science to help, and uses three questions to turn attention towards what can be done in the present by the pupils themselves. Another pupil (George) now suggests picking up litter and putting it in the bin, before immediately telling the class a story about cooperation emerging out of conflict.

This brief interaction lasted 1 minute 33 seconds, out of a total of 6 hours and 11 minutes of video data (so represents a tiny fraction of the data analysed; i.e., 0.42%). The PAF can be used to identify significant elements in play during interactions like this in classrooms (for example, in the 18 lines of interaction analysed above there were 7 means (i.e., a person or a thing), 21 instances where someone informs someone else of something, 3 actions, 2 uses of space/time, 5 questions of different sorts, 1 example of something being repeated, and 2 instances where the context becomes significant; see Riordan et al., 2021). There is no suggestion here that participants would identify elements of the PAF as they teach or learn. I argue that pedagogy analysis, by teachers watching video of their own practice or that of another teacher, or by pedagogy researchers, does sometimes get so complicated in lessons that the PAF can be of use (see Riordan et al., 2021). I also acknowledge that the transcript above can be analysed in other, perhaps less pedantic ways.

Sustainability pedagogy as ethical pedagogy

In this section, drawing on the short interaction above, I demonstrate how sustainability pedagogy can be considered as an example of ethical pedagogy. I understand ethical pedagogy to be where interactions in the classroom have an ethical focus. For example, ethical pedagogy could be concerned with teaching about an ethical topic (e.g., colonisation), or teaching in an ethical way (e.g., avoiding colonising pedagogy), or teaching using ethically sourced resources (e.g., using fair trade resources). Following Beach and Eriksson (2010), I draw on a philosophical overview of ethical positions from Flinders (1992, p. 101) who proposed that; utilitarian ethics focusses on the consequences of moral reasoning, deontological ethics on conformity to rules, relational ethics on respect for others, and ecological ethics on the interdependence of people (and of people with nature). Any of these ethical positions can underpin ethical pedagogy.

Ms Ivy's question could be understood as reflecting utilitarian ethics (i.e., focusing on the moral consequences of the actions the participants take to 'make a difference'). Oliver's solutions and future plans may not have many ethical implications, but George's litter picking idea can be related to deontological ethics (conformity to the rules of responsible waste management), and his final point about the football match during World War I, may display relational ethics in that it demonstrates how enemies can respect each other, and in the way adversaries can sometimes change the way they perceive each other. Consequently, ethical pedagogy can be concerned with the analysis of interactions in a classroom from a particular ethical position (e.g., using utilitarian, deontological, relational, or ecological ethics), and participants in a classroom may take different ethical positions. Hence, sustainability pedagogy must seek to understand and explain the diverse ethical assumptions of participants as they interact with each other and with things.

To develop this thinking further about the ethical positioning of sustainability concerns in a classroom situation, I present an imaginary interaction set in a UK primary school classroom between Ava and Leo (both aged 8).

Ava sees Leo putting paper in the general waste rather than in the recycle bin. Getting a little angry she tells him, 'That's wasteful'. Ava knows the school bin system and that Leo has done something wrong, and she wants him to know that she knows he has done something wrong with the hope that next time Leo will use the correct bin. What Ava does not know is that the school caretaker is Leo's father, and he has told Leo that it doesn't matter what bin anyone uses, as he puts all the rubbish into the same big bin outside. Leo resents Ava's unsolicited advice, after all he has put the paper in a bin. He feels annoyed and responds, 'Mind your own business!'. A conflict is potentially brewing.

Taking a utilitarian ethical stance – Ava reprimanding Leo for having thrown paper in the wrong bin – might have the positive consequence, from Ava's perspective, of appropriate recycling behaviour in future, as Leo may be fearful of further penalties. Next, from the perspective of deontological ethics, in some classrooms berating Leo in such circumstances would be perceived by pupils to be morally justified. Ava seems to think her intervention is fair. However, drawing on relational ethics, Leo might consider that Ava's words and tone are not respectful. Regarding ecological ethics, Leo's bin use could be understood by him (but not by Ava) as a silent protest at what he perceives as Ava's domineering behaviour from before this present interaction. Although hypothetical, this vignette shows how a variety of ethical positions can influence pedagogical problems (Riordan et al., 2021). For example, the place *where* an interaction between someone and an object occurs can have ethical implications. So, paper dropped in one bin (the general waste) could be considered wrong from Ava's perspective, but the same paper dropped in a correct bin (the recycling) is fine for her. In contrast Leo knows that his Dad (the caretaker) puts all the waste in the same bin outside anyway. So Leo knows Ava's fastidiousness is pointless, whereas Ava is unaware. The vignette highlights how participants can sometimes interpret the same classroom interaction that has ethical implications in different ways, that any participant in classroom pedagogy can shape the physical and cultural contexts over time, and that the aforementioned pedagogy is itself moulded by these dynamic contexts.

Developing sustainability pedagogy in practice

Given the importance of managing ethical relationships within sustainability pedagogy, it is important to give new teachers the opportunities to engage with classroom video of real interactions between pupils, and between pupils and teachers. Teachers contribute to the moral education of the learners in their care, so the ability of teachers to understand and explain ethical positions is important. Furthermore, in my experience from when I was a school teacher with teenagers and now as a teacher of adults, many learners can spot incoherence between what a teacher says and what they do, and this hypocrisy can seriously undermine the credibility of a teacher in a classroom with sometimes serious consequences (e.g., disintegrating behaviour for learning). Some teachers are self-aware when they make such mistakes (and may or may not address these with the learners), but others may not even know what they have done. I think there should be a place in ECT coaching, and in university-based Initial Teacher Education, for detailed untangling of complicated interactions that happen in real classrooms (and a method like the PAF may help with those processes).

I am currently working with school teachers on a new research project exploring how such classroom video analysis methods could be used in Early Career Teacher coaching. This project could reveal examples of coaches helping ECTs with video analysis of ethical pedagogy. I imagine an Early Career Teacher and their coach watching the video of the interaction of Ms Ivy with Oliver and George, and untangling what is happening together, evaluating the pedagogy, discussing critically how a passage of the same lesson is used in a peer-reviewed academic journal article (e.g., Riordan et al., 2021), analysing a two-minute video clip of the ECT's teaching (and/or that of their coach), and then using these learning experiences to plan subtle tweaks in the ECT's own practice during the next lesson. This might lead to richer learning experiences during Early Career Teacher coaching sessions than a general conversation based on a question like, 'How are things going?'

Discussion

This chapter has explored some ways in which sustainability pedagogy can be understood, illustrating how analysis of real classroom interactions can contribute to Initial Teacher Education regarding sustainability. The following discussion explores five questions concerning sustainability pedagogy.

(1) How can/should sustainability be taught in the classroom? What pedagogy helps (or hinders) children's varied understandings of sustainability? Managing ethical pedagogy, like sustainability pedagogy, is sometimes difficult for teachers, and novice–expert strategic dialogue can struggle for want of a shared vocabulary, so approaches supporting analysis of ethical pedagogy could be useful for school coaches, teacher educators, and pedagogy researchers. The PAF analysis above, discusses a videoed example of complicated sustainability pedagogy to show how this approach can provide such a shared vocabulary.

(2) How do real teachers teach about ethical and unethical behaviour, like sustainability and unsustainable behaviours, in an ethical way? For example, educating about sustainability is sometimes combined by teachers with pedagogy seeking behaviour change, so ideological tension between knowledge transfer intention and the societal change intention can make such pedagogy difficult to lead in the classroom (and to analyse). Regarding sustainability, the PAF can be used to identifying the tactical and strategic ways in which resources are used in complicated pedagogical encounters). I think other types of ethical pedagogy also merit investigation, and that similarities and differences between such pedagogies could be interesting (hence my next research studies about 'decolonising pedagogy' and about 'disagreement (including conflict) in classrooms', two interrelated topics).

(3) How do children understand a concept like 'sustainability' – both before and after teaching? Though some literature exists on sustainability misconceptions in higher education (e.g., Leal Filho, 2000), there seems to be little research regarding children's sustainability misconceptions, nor much about conceptual change pedagogy in this area. If the 'misconception movement' (diSessa, 2006) is anything to go by, children's thinking about sustainability may turn out to be wildly different from that of adults, and this may have profound implications for sustainability pedagogy and sustainability theory more widely. Video-based pedagogy analysis research, using the PAF, could be a fruitful way of exploring this. For example, in the lesson analysed above, I also have 'pupil group verbal protocol' video data on that lesson, where a small group of pupils from the lesson later watched their lesson back and 'thought aloud together'.

Triangulating research methods like this can reveal evidence of conceptual change and of the pedagogy that may have brought about those changes (see Riordan et al., 2021).

(4) How should a classroom teacher accommodate (or not) learners who hold ethical assumptions differing to that which underpins the lesson? How should a participant whose personal ethical system does not correspond to the one used in their lesson manage the personal cognitive and affective challenges this entails? For example, as mentioned earlier, I recorded the data in this chapter in a UK faith school. Arguably, parents, or carers of pupils at faith schools in the UK indicate their openness to religious education by sending their child to such a school. Nevertheless, faith communities in the UK are generally diverse. Many families whose children attend faith schools are not themselves religious. Many teachers who work in UK faith schools identify as members of that faith community, but others are part of different faith communities, and some not part of any faith group. A further complication is that a competitive system for school places in some parts of the UK means some parents feel under pressure to send their child to a faith school, may not be comfortable with the schools' ethical system, and/or may or may not be open about their discomfort with the school for many reasons (i.e., deception is sometimes used). So, a wide variety of ethical assumptions are to be expected among teachers, parents (including carers), pupils, and indeed researchers.

(5) What sorts of pedagogical dilemmas are implicated in sustainability pedagogy? In Riordan et al. (2021) we argued that pedagogical problems are often messy, and heuristic solutions (defined by Berlyne, Vinacke, and Sternberg (2008), as informal, intuitive, and speculative procedures that lead to a solution in some cases but not in others) are more appropriate than algorithmic ones.

An algorithm is certain to lead to a solution (if followed correctly), but a heuristic is not. Furthermore, pedagogical problems are often 'wicked' and 'ill-structured' (Reed, 2016). Pedagogical problems in large classes often occur simultaneously and blend into each other (sometimes termed 'cyclical' in the literature, as one problem becomes part of the context of the next: Pretz, Naples, and Sternberg, 2003). Hence there can never be a recipe (i.e., an algorithm) for how a teacher should address a difficult pedagogical ethical question like those raised in this discussion so far. What school coaches and other teacher educators can do is educate new teachers in ethical pedagogy analysis, perhaps using an exemplar like that analysed in this chapter (with the accompanying video data) and ones from their own school context, to help new colleagues make pedagogical decisions in their own classrooms. Furthermore, new teachers need to be educated so they can engage critically with 'educational gurus' who offer simple instructions for how to solve complex pedagogical problems. I would argue that a teacher in a particular classroom cannot teach like anyone else (even like those who consider themselves 'classroom champions') as the pedagogical problems each teacher faces will be unique to the changing and complex context a teacher works in. Beware pedagogical snake oil.

Conclusion

What then does this chapter tell us about sustainability pedagogy and about how pedagogy research can contribute to Initial Teacher Education and Early Career Teacher coaching regarding sustainability pedagogy? First, sustainability pedagogy in real classrooms can get complicated very quickly and teachers (e.g., student, early career, or experienced), and teacher educators (e.g., school coaches, university-based teacher educators, etc.), may sometimes find approaches like the PAF useful. How exactly PAF might be useful requires further research, and that study is underway. Second, I think access to video examples from real classrooms of interactions like this to be used for educational purposes is potentially valuable for learners and educators. Such materials must be collected and stored in ethical ways of course. For example, I have ethical permission from participants and the university ethics committee to keep such video materials in the long term in a secure online repository at the UK Data Service called 'ReShare' for archival, research, and teaching purposes (see Riordan et al., 2022 and the data availability statement at the end).

Finally, bridging the gap between theoretical writing about sustainability and 'what I can/should do in my classroom regarding sustainability' may be challenging for some teachers, particularly near the start of their careers, so pedagogy research that links real classroom examples with sustainability theory may be helpful. To illustrate, here are two questions new teachers could explore, basing their thinking on video of real classroom interactions like those in the clip analysed above. How do real classroom teachers encourage thinking about sustainability and guide reasoning towards practical solutions? Also, what is reasonable to expect of primary school pupils regarding potentially abstract notions of sustainability? Incidentally, real plastic-eating robot turtles were already used in the ocean at the time this lesson was recorded, but these participants seemed to be unaware of that.

To finish, I return briefly to Ava and Leo's altercation. Ava could reflect on ethical pedagogy which might have made achievement of her intended purpose more likely. To illustrate, Ava could have refrained from commenting on Leo's misdemeanour and instead lavished praise publicly on the next person to use the recycle bin correctly. Thereby eschewing giving attention to Leo's misdemeanour, avoiding annoying him through giving unsolicited advice, and demonstrating to Leo the sort of ethical behaviour that does result in attention (i.e., a reward). Tactical use of praise like this will be familiar to teachers reading this and was used frequently by the teacher in the lesson studied. There is no guarantee that this approach will work with Leo of course – he might be enjoying throwing paper in the wrong bin.

Data availability

The data that support this chapter are available at http://doi.org/10.5255/UKDA-SN-854915.

Acknowledgements

I thank the school, the teacher, the TAs, all the pupils, DS, GH, JR, CR, and HR. I thank my colleagues in the NICER team at Canterbury Christ Church University. This work was supported by the Templeton World Charity Foundation (grant reference TWCF0375).

References

Alexander, R. (2008) *Essays on Pedagogy*. Abingdon: Routledge.

Beach, D. and Eriksson, A. (2010) 'The relationship between ethical positions and methodological approaches: A Scandinavian perspective', *Ethnography and Education,* 5(2), pp. 129–142. doi: 10.1080/17457823.2010.493393

Berlyne, D.E., Vinacke, W.E. and Sternberg, R.J. (2008) *Thought.* Available at: https://academic.eb.com/levels/collegiate/article/thought/108663 (Accessed: 19 May 2021).

Black, P. and William, D. (2018) 'Classroom assessment and pedagogy', *Assessment in Education: Principles, Policy & Practice,* 25(6), pp. 551–575. doi: 10.1080/0969594X.2018.1441807

DfE (2022) *Sustainability and climate change: a strategy for the education and children's services systems.* Available at: www.gov.uk/government/publications/sustainability-and-climate-change-strategy/sustainability-and-climate-change-a-strategy-for-the-education-and-childrens-services-systems" \t "_blank (Accessed: 17/12/22).

DfE (2014) *The national curriculum in England framework document.* Available at: https://assets.publishing.service.gov.uk/government/uploads/system/uploads/attachment_data/file/381344/Master_final_national_curriculum_28_Nov.pdf" \t "_blank (Accessed: 9/7/22).

diSessa, A. (2006) 'A history of conceptual change research: threads and fault lines', in Sawyer, K. (ed.) *The Cambridge Handbook of the Learning Sciences.* Cambridge: Cambridge University Press, pp. 265–282.

Flinders, D.J. (1992) 'In search of ethical guidance: Constructing a basis for dialogue', *Qualitative Studies in Education,* 5(2), pp. 101–115. doi: 10.1080/0951839920050202

Hallam, S. and Ireson, J. (1999) 'Pedagogy in the secondary school', in Mortimore, P. (ed.) *Understanding Pedagogy and Its Impact on Learning* London: Paul Chapman, pp. 68–97.

Hardman, M., Riordan, J. and Hetherington, L. (2022) ' A material-dialogic perspective on powerful knowledge and matter within a science classroom', in Hudson, B., Stolare, M., Gericke, N. and Olin-Scheller, C. (eds.) *Powerful knowledge and epistemic quality across school subjects.* London: Bloomsbury Academic, pp. 157–175.

Kopnina, H. and Meijers, F. (2014) 'Education for sustainable development (ESD): Exploring theoretical and practical challenges', *International Journal of Sustainability in Higher Education,* 15(2), pp. 188–207. doi: 10.1108/IJSHE-07-2012-0059

Leal Filho, W. (2000) 'Dealing with misconceptions on the concept of sustainability', *International Journal of Sustainability in Higher Education,* 1(1), pp. 9–19. doi: 10.1108/1467630010307066

Pörtner, H., Roberts, D.C., Poloczanska, E.S., Mintenbeck, K., Tignor, M., Alegría, A., Craig, M., Langsdorf, S., Löschke, S., Möller, V. and Okem, A. (2022) 'IPCC Summary for Policymakers', in Pörtner, H.-., Roberts, D.C., Tignor, M., Poloczanska, E.S., Mintenbeck, K., Alegría, A., Craig, M., Langsdorf, S., Löschke, S., Möller, V., Okem, A. and Rama, B. (eds.) *Climate Change 2022: Impacts, Adaptation, and Vulnerability. Contribution of Working Group II to the Sixth Assessment Report of the Intergovernmental Panel on Climate Change* Cambridge: Cambridge University Press.

Pretz, J.E., Naples, A.J. and Sternberg, R.J. (2003) 'Recognizing, defining, and representing problems', in Davidson, J. and Sternberg, R. (eds.) *The Psychology of Problem Solving*, pp. 3–30.

Reed, S.K. (2016) 'The structure of ill-structured (and well-structured) problems revisited', *Educational Psychology Review,* 28(4), pp. 691–716. doi: 10.1007/s10648-015-9343-1

Riordan, J.P., Hardman, M. and Cumbers, D. (2021) 'Pedagogy Analysis Framework: a video-based tool for combining teacher, pupil and researcher perspectives', *Research in Science & Technological Education.* doi: 10.1080/02635143.2021.1972960

Riordan, J.P., Revell, L., Bowie, B., Woolley, M., Hulbert, S. and Thomas, C. (2022) *Video-Based Study of Classroom Pedagogy, 2019–2021. [Data Collection].* Colchester, Essex: UK Data Service.

Riordan, J.P., Revell, L., Bowie, B., Woolley, M., Hulbert, S. and Thomas, C. (2021) 'Video-based grounded theory study of primary classroom strategy: using the extended Pedagogy Analysis Framework to understand and explain problem-solving when science and RE topics interact', *Research in Science & Technological Education.* doi: 10.1080/02635143.2021.2001450

Riordan, J. (2022) 'A method and framework for video-based pedagogy analysis', *Research in Science & Technological Education,* 40(1), pp. 53–75. doi: 10.1080/02635143.2020.1776243

Salas-Zapata, W.A., Ríos-Osorio, L.A. and Cardona-Arias, J.A. (2018) 'Knowledge, Attitudes and Practices of Sustainability: Systematic Review 1990–2016', *Journal of Teacher Education for Sustainability,* 20(1), pp. 46–63. doi: 10.2478/jtes-2018-0003

Taylor, L. and Dionne, J.P. (2000) 'Accessing problem-solving strategy knowledge: The complementary use of concurrent verbal protocols and retrospective debriefing', *Journal of Educational Psychology,* 92(3), pp. 413–425. doi: 10.1037//0022-0663.92.3.413

White, J. (2008) 'Sustainable pedagogy: A research narrative about performativity, teachers and possibility', *Transnational Curriculum Inquiry,* 5(1), pp. 1–14.

The tangled bank

Victoria Field

Have you found it interesting to contemplate the tangled bank on campus?

Have you seen the ox eye daisies, ladies bedstraw, common toad flax, cowslips and tall grasses softening the straight lines and corners of our built environment?

When did you last listen to the birds singing on the bushes?

Have you seen various insects flitting about? Have you seen any insects?

When did you last have your hands in the damp earth through which the worms are crawling?

Whose laws are you obeying when you forget to honour these elaborately constructed forms?

When did you last acknowledge our beautiful differences and dependencies? How?

DOI: 10.4324/9781003286516-15

Part 3

(Higher) education as if the world mattered

The final part of this book asks difficult questions about past educational philosophy and structures and seeks to re-imagine what a 'good' education in a fragile world might be like if universities were to take the challenge of sustainability seriously. Wilding is offered as one potential approach to challenge the neoliberal forces shaping the HE sector within the contemporary context. Transition education, involving radical re-structuring of societal conditions and educational processes is offered as another alternative to counter existing dysfunctional systems. Finally, focusing on how universities could put their transformational intent into practice, a provocative 'Trojan Horse' model of ESD is offered.

DOI: 10.4324/9781003286516-16

11 Wilding higher education

From monoculture to messy margins

Alan Bainbridge

A vignette of a tamed higher education sometime in the future

Alan had slept well and woke up looking forward to working through today's Level 11c Edu-success Challenges. Last night he completed the preparation material (Level 11b) for Module x2913 (Integrated Global Education) that had been posted by Dr Kemp on the Digi-flex virtual learning platform. The learning tasks had not been too difficult; fortunately, Dr Kemp had provided a 200-word summary of Winchester's internationally validated seminal work, 'Educational Solutions for Sustainable Futures'. There were also prompt questions with 'clues to the answers' icons to click on if you got stuck. By selecting the 'Continuous Improvement Feedback' option (CIF), Alan was able to monitor his learning gain in real-time compared to all other UK Education Studies students and was pleased that his position in the top ten percentile of the league table had been maintained. Not bad for someone at the Ambition Trust University (ATU), currently languishing in the depths of the Aggregated Media University league table.

Despite winning the Excellent Student Services award for five years and being the only higher education institution to gain a 'Love the Planet' kitemark since its inception, ATU was failing to attract enough students to remain financially secure. Talks had already begun to reduce ATU staff numbers and merge with a solvent and expanding Hilton Group Hyper-university. The academic-to-student ratio would rise from 1:150 to 1:250, but the Hilton's superior multi-verse online provision and prestigious 10-star research profile would still ensure their top 20 position could be maintained, if not improved.

Anyway, Alan wasn't concerned about ATU's future. He was on a roll and motivated by last night's micro-formative feedback. Choosing to walk through the park, he packed and set off to the campus wearing the brilliant Digi-flex algorithmic Thought Capture headset. Thoughts could be detected, standardized across huge data sets, and used to provide a bespoke educational experience that would maximize learning capacity (Digi-flex claimed a 2-semester learning gain for 99% of regular users). Alan selected the option to listen to a longer version that provided a precis of each chapter of 'Educational Solutions for Sustainable Futures' – read by Dr Kemp. By the time he reached the lecture hall, the CIF beeped an alert. Checking the reading, Alan could see that he had begun to slip

DOI: 10.4324/9781003286516-17

down the UK Education Studies student league table. This had happened before; something must have distracted his learning. Alan didn't panic and settled into the lecture, opting to use the Digi-flex learning alignment function that identified gaps in his learning, adjusting the pre-recorded lecture to ensure an optimal learning experience. At the end of the lecture Alan looked around at his peers and wondered what they had learned and where they might be in the league table. He was happy enough though, the CIF reported an immediate learning gain – he had leapt into the top 5 percentile. Meanwhile, the ATU website reported the successful merger with the Hilton Group Hyper-university with a new motto scrolling across the Home Page – 'Y reach for the stars when U can become 1 at a Hilton Hyper-university?!'. (After Wallin, 2022).

Introduction

This chapter notes that contemporary higher education is potentially a site of significant innovation, and therefore an important driver for a more hopeful sustainable future. Despite this, a provocation is offered arguing that the potential to promote sustainable living in an increasingly fragile world has been threatened by the 'taming' of higher education, and the case for promoting a 'wild' education will be made. The vignette above (and the one that finishes this chapter) indicate some aspects of the wild/tame metaphor, and although fictional may no doubt resonate with the experience of many readers. The use of these metaphors, although a deliberate attempt to draw attention, is not offered as a gimmick, but rather as an acknowledgment that there are useful links between the ecological metaphor of tame/wild and the language and principles of good education. Central to this argument is the assumption that the activities and processes of ecological wildness and education operate according to similar principles and logic. Both are complex, open to serendipity, riven with tension and held in fine balance and although often stable, subject to considerable transformation. The idea of 'wildness', which will be explored later, is argued to be relevant to all phases of educational experience, but first, the case will be made for its resonance and applicability to higher education as an important site of individual and cultural innovation and transformation.

The transformative potential of higher education

Higher education is of particular interest due to its significant role to influence the development of potentially transformative aspects of social, economic, and cultural life (Moscardini, Strachan & Vlasova, 2020). Drawing on the ecological metaphor, this influential positioning can be used to frame higher education as a 'keystone species', where the actions and outcomes of practice can both determine, or indicate the 'health' of the structure and functioning of the wider educational, economic, and ecological community. Sterling (2021) also conceives higher education from an ecological perspective, as an innovative, or adaptive learning institution that, in the right conditions, co-evolves with the local and wider community.

Even at this early stage, it is important not to set higher education up as a neutral panacea for the current global environmental and non-environmental poly-crisis. Stein (2019) recognizes the invasive and exploitative global impact of a 'European-influenced' higher education, calling for colleges and universities to decolonise and step outside of the hopes and promises they currently offer. Therefore, a good higher education for a sustainable planet must take advantage of its 'keystone' positioning and go beyond existing modes of thinking that have, it could be argued, been partly responsible for, and failed to halt planetary crises of climate change, biodiversity collapse, and resource depletion. Indeed, a 'good, or wild' higher education can be positioned as a 'red-flag species' – on the edge of extinction, collapsing under the demands of neoliberal accountability – providing a warning across the education sector of what might be lost as it is tamed, potentially leading to its demise.

From this context, drawing on Ingold (2023), I suggest that widespread contemporary (even global) education policy, influenced by positivist enlightenment and neoliberal assumptions, has led to a denuded and tamed conception of formal education that focuses on predictive and directional *intention* to deliver a particular curriculum in a particular manner. This can be observed through the dominance of subject silos and simplistic notions of cause-and-effect educational thinking where progress from one 'level' to the next is conceived, with the 'appropriate' support and feedback, as a relatively straightforward progression towards individual excellence. Such ideas are less helpful when considering the relationship between education and sustainability, where it will be shown later that 'wild epistemologies' (Rowlandson, 2020) are required to direct *attention* to embrace messiness, confronting difficult questions, encouraging debate, and being able to live with uncertainty and dissent (Ingold, 2023).

A wild higher education, due to its 'keystone' influence, will be shown to be found in a reparative shift from education that is not so much about positioning students *in* the world, compared to encouraging ways *of* psychologically encountering the other in the world (Skea and Fulford, 2021), and the rejection of the dominance of *intentional* models of education. Alternatively, what is offered is the possibility of education as a weak and uncertain process (Biesta, 2013), at odds with contemporary neoliberal assumptions of simplistic cause-and-effect, and the need to engage with complexity and the possibility of not knowing (Bainbridge, 2019). It will be an education that can sustain itself, resisting continual attempts to be tamed, fulfilling Barnett's (2018) call for universities to be the sites of revolutionary thought and action; and an education that recognizes the *response-ability* of learning to live together with more-than-human others in a world that supports mutual flourishing.

This is the site of higher education's transformative potential, where new thinking about living sustainably in a finite world can emerge. It is an educational 'keystone' space where difficult questions can be asked, anxiety can be expected and held, and the desires of the individual and non-human other can be thought about. It is not a place of replication, model answers, or indeed 'excellence', but a 'taskscape' (Ingold, 1993) where knowledge and wisdom are embedded in listening and paying attention to the planetary processes that support the life of all those with whom we share the planet. At the heart of wilding higher education is the principle of being aware and attentive to the human relationship with the wider external world, to its complexity and serendipity, and a psychological openness to embrace multiple means of exploration, the outcomes of which might not be predictable. It will be here that contemporary wicked problems can be encountered, difficult decisions can be made, perhaps even acknowledgment of the negative consequences of human activity. A good education in a world made fragile by the activity of humans will be an education where the anxiety of learners can be held by teachers/ lecturers (and parents), enabling them to be attentive in the presence of difficult knowledge, complex ideas and ultimately to consider their response to the more-than-human world.

Wilding education

It is important to start with the rationale for using the term 'wilding', rather than the more commonly used 're-wilding' that makes assumptions of a return to what an ecosystem used to be like and would include the re-introduction of species no longer present. In contrast, wilding does not involve a return to the past but recognizes the importance of using natural processes to benefit the environment, usually to stabilise an ecosystem by increasing biodiversity (complexity). The parallel educational argument presented here is not for a return to 'how life/education used to be in the old days', but how principles of being 'wild' can inform the development of an education that is sustainable and not itself on the edge of extinction; while also offering a way of being human and learning to live harmoniously with the more-than-human other.

Despite playfully engaging with ecological metaphors and principles of planetary interconnectedness, it must be made clear that good education is not necessarily predicated on acts of physical contact with the natural world, but instead through an awareness of the psychological stance required to live within complex interdependent relationships between human and non-human others. Opportunities for wild education, leading to a deeper sense of what our roles and responsibilities to others and the planet might be, can be found in a heightened awareness of, and paying attention to, the continually changing local and global worlds. This represents a shift in focus that can also be defined in terms of moving from individual ego-logical thinking (intentional), to eco-logical (attentional) considerations of who I might be, and what my responsibilities might be in the presence of an other (Bainbridge and Del Negro, 2020).

The discussion developed here acknowledges the complexity of ecological interrelations that, although neutral in sum, involve a balance between winners and losers, life and death, disease and health. A deeper awareness of ecological processes can therefore bring us into closer contact to experiences that can oscillate between both beauty and darkness, joy and fear, or even success and failure. The tamed education outlined in the first vignette suggests a substantive move away from complexity, serendipity, and civic responsibility, gravitating instead towards a global curriculum monoculture, quasi-competitive league tables, and the certainty of progress towards excellence. To consider what other epistemological possibilities a wild education might offer, I shall try to imagine and present a mind not encumbered by modern-day technicist thinking. Arguably, this shift is particularly significant within the context of a potentially transformative higher education environment, where innovation and critical thinking should be thriving. To highlight what I mean by stepping out of modern-day technicist thinking and embracing a more ecological mindset, I shall first consider human encounters with nature that have the potential to disrupt and hold our attention in different ways. Although usually experienced as occasional interruptions on the margins of the everyday, I will suggest that in the shift from ego-logic intention to eco-logic attention, awe and wonder, complexity and uncertainty are encountered and something new has the potential to emerge.

Encounters with nature can offer moments for subtle shifts from an ego-logical to eco-logical mindset. The young child's fascination with the tiny world of an ant crossing a path that momentarily shifts their attention, draws them into a world of 'ant-ness', releasing them from their own wants and desires. An adolescent immersed in the world of social media can be sufficiently moved by images of environmental degradation, to put aside their own comfort and safety to protest and risk arrest to draw attention to wider ecological planetary needs. The lifelong birdwatcher searching the skies for the return or departure of migrating birds focuses less on work schedules and targets, instead paying careful attention to the impact of changing seasons and climate breakdown. Or the everyday experience of recognizing an 'old friend' when crossing the path of an urban fox or finding yourself looking into the eyes of a robin feeding in the garden. A sudden and unexpected encounter with the natural world can reveal more about the relationship between humans and nature than is first obvious, for example, children in a classroom disturbed by the sudden entry of a noisy wasp are drawn away from the trigonometry test at the end of the week and suddenly are gripped with fear at the potential pain of a sudden stinging attack. Of course, there will be others who sense the wasp's own fear of being 'out-of-place' and they will rush with up-turned drinking glasses to rescue the unfortunate insect.

In the examples above, senses are momentarily cast back to an ancient relationship, as the natural other is rapidly decoded as friend or foe, food or threat: and at this moment colour, sound, movement, and smell become more defined and possibly overpowering. This is where Abram (1997) argues that we become truly human, for just as our sense of self is experienced through how others recognise and respond to us, so we become human when we are aware of being in the presence of the more-than-human other. This is a state Abram refers to where mind and world are not separate, and the biosphere is experienced from within. These are intensely attentive experiences that, when stumbled upon, can become educational moments where individual needs and wants may be put aside in consideration of a more-than-human other.

Ironically, the products of a human world, socially and culturally constructed, have led to the contemporary human condition, of living in the presence of nature, but largely unaware of the rich aliveness and potential such a relationship has to offer. But to be aware of this also highlights the emotional uncertainty (awe and wonder, fear and beauty) of the reality that humans may only be comparatively short-term actors in the long-life of a continuously evolving planet. I have argued elsewhere (Bainbridge, 2019) that for reasons possibly unknowable, human ego-logical thinking is potentially an avoidance of the emotional uncertainty inherent within eco-logical states. Additionally, that ego-logical thinking is maintained via unconscious ego defences to protect humans from the potentially overwhelming anxiety of not knowing in the face of uncertainty and complexity. Consequently, any discussion on the wilding of educational experience will require consideration of psychological processes, particularly how uncertainty and complexity can be expected and experienced rather than avoided.

Modernity and all its benefits, including education, have therefore come at a price – that is, the ever-increasing attempts to control and manage represent a move towards tame ways of thinking that focus on ego intention and away from eco attentiveness and the needs of the other. Indeed, an understanding of an ecologically interconnected world would lead to an acknowledgement that human-instigated actions of construction and control have a negative impact on the more-than-human world. It could be argued that the human condition – dilemma, even – is to be caught between a desire to engage in attempting to tame and manage the planet's resources. Meanwhile such actions may be responsible for removing from awareness the reality of the wildness of an interconnected existence within a complex ecological world.

On one level, living with the more-than-human natural world is a very ordinary everyday experience – all but a very few of us will go through a day where we do not see sunlight, feel the breeze or see plants and animals. Wild encounters such as these can also offer moments of awe and wonder, of comfort and discomfort, when our attention slips away from ego desires, allowing the present to be transcendent and relational with an other in the here and now. But it is not just being in the presence of a natural other that frames such everyday moments as potentially educational. Rather, a wild educational encounter is to engage in a less observational but more sensitive, relational, and responsive manner. Returning to Skea and Fulford's (2021) distinction between education 'in' or 'of' the outdoors, the wilding of education is presented as less dependent on being physically present *in,* and more contingent on different psychological ways *of* encountering and paying attention to the world. It is the application of this psychological re-direction; from ego to eco, from control to uncertainty, and from cause-and-effect to complexity, through which a tamed higher education may now become wild and more likely to provide a 'good education for a fragile world'.

Wilding higher education

At the heart of the idea of wilding higher education is the potential confusion outlined above, that physical contact with nature is prerequisite to wildness. The proposition presented here is psychological rather than physical: that wild education is supported by eco-logical assumptions, and not defensive ego-logic thinking. The intimate and immediate nature of the non-human/human relationship becomes clear when, for example, we consider the impact of human activity on carbon dioxide levels leading to climate breakdown, and how a microscopic coronavirus has had a global impact on human activities. This issue is not so much one of human disconnection from the natural world, but rather in relation to conscious, or unconscious processes, to be aware of alternative ways of thinking that would include considering what it means to be dependent on maintaining planetary interconnectedness.

The disconnection/unawareness distinction can be seen in an education situated 'in' the world that makes demands to facilitate the delivery of a particular curriculum and, for assessment to lead to increased success – even excellence. In this context, the external world is approached from a one-directional position of intention, with a focus on what can be provided and what particular individual outcomes might be. In an education 'of' the world, the learner is required to suspend judgment, engage and pay attention to the human/more-than-human other within an open reciprocal relationship. Such a stance requires those involved in the educational encounter to inhabit a less defended position, and to be more open to taking risks; including being able to sit with not knowing, offering opportunities for new thinking where each can account for their own actions. Some readers will be fearful that what is suggested may lead to uncontrolled *laisse faire* anarchy, where students are given complex real-world problems to solve at their own whim and fancy. This, though, is not the intention – staying with the ecological metaphor, the wilding of higher education does not require every aspect (management, teaching and research) to be opened up to serendipity and complex uncertainty. Sudden and wholescale change in ecosystems and education settings is likely to be catastrophic and fatal. Just as ecological principles would support the wilding of small areas, for example; not cultivating the edges of arable crops, leaving some hills free from grazing, or not mowing during May, so the wilding of higher education can be achieved by similar activity around the margins.

Lists and concrete examples are always risky and insufficient – indeed a wild education would be skeptical of 'scaling-up' local institutional processes and practices, suggesting that this itself is an act of taming – but it is acknowledged that readers would welcome something a little more tangible than 'encountering complexity'. Therefore, to ground the ideas previously discussed, a wild higher education *may* feature the following:

- A less structured curriculum, perhaps providing opportunities for interdisciplinary study and research.
- A variety of assessment techniques negotiated with students leading to multiple outcomes where the focus is not so much on comparing student against student but developing *response-ability*.
- Provide opportunities for real-life problem solving, including socio-economic thinking systems to highlight complexity.
- Engage with the local community to identify areas of concern and encourage activism and participation.
- In a profession dominated by measures of excellence and league tables, develop opportunities for collaboration not competition.
- Challenge orthodoxy in management and academic thinking – make space for creativity, spirituality and Global South world views.
- All managerial, pedagogic and research decisions need to be grounded in an understanding of an utter dependency on maintaining a healthy planet.

Conclusion: higher education as sustainability

It has been argued that modern humans have lost or are losing their sensitivity, awareness, and responsibility towards all that is not human. The sustainable, and potentially paradoxical challenge for education is to enable the benefits of modernity to be enjoyed, while also experiencing a meaningful awareness of, and to act in such a way as to not harm, the relationship between the human and more-than-human other. An education such as this requires an ability to be attentive while sitting with uncertainty and complexity, rejecting the hegemonic influence of simplistic cause-and-effect models of learning.

A wild education does not have to involve lessons on nature, sustainable behaviours, or a desire to return to past pre-industrial times. Instead focusing on psychological factors, providing opportunities to disrupt thinking, without an expectation of quickness or certitude, waiting for learners to consider their response, in light of their desires in relation to the needs of others, including the more-than-human world. A wild education will be able to live with uncertainty and discomfort. Abram (1997) calls for us to become apprentices to our local environment – to look and listen closely to what is around us, not to name or control, but simply to be able to come into its presence from a sensitive awareness of equality and dependency. Equally, a wild higher education requiring us to accept that we remain apprentices, not experts of the worlds we inhabit, willing to pay attention and consider how the other is encountered ,will therefore represent education *as* sustainability.

A wild higher education sometime in the future

Alan had slept OK and woke up feeling slightly apprehensive about starting the Year 3 Integrated Global Education module today. He had read the chapters Dr Kemp suggested but struggled to make sense of how Fanon's 'Wretched of the Earth' offered any insight into education policy. The virtual learning platform had a discussion room and it seemed that he was not alone: most of his fellow students seemed equally baffled, but Mila, the exchange student, loved it and had shared a link to a video that he decided to watch on the bus. Alan was still uneasy, though: he usually got at least 65 per cent for his essays, but he felt this time he might scrape a 40 per cent; he got a 45 per cent in Year 1 and didn't want to repeat that experience. What was Dr Kemp thinking!? Probably promoting their new 'big idea' to a captive audience.

Alan was quite enjoying his time at Christminster University. It was not one of those famous universities you see on quiz shows, but he felt well-supported, especially early on when he needed a little more help to write in an academic manner. He really enjoyed the sports and free film screenings and had even led some protests against a road being built through a wood he liked to walk in at the weekend. He felt proud to wear the university hoody with the 'Love the Planet' award logo – especially as the university sustainability team used his protest in their winning submission. Next year, the university was planning to use the government's 'Universities for All' funding opportunity to offer free 'taster' courses to anyone who wanted to attend – he thought, if he was still in town, he would go along. Maybe one day he could be a lecturer?

By the time the bus arrived near the campus, he had watched the video and had a better grasp of Fanon's ideas of how power, violence, and education can be connected. Alan happened to meet Mila by the bus stop, and they immediately began to talk, far too fast, about Fanon and what Dr Kemp might have to say. Anyway, they planned to ask a 'demon' question in that awkward silent bit during a lecture when the students were invited to pose any questions. Sitting in the lecture theatre, Alan took out his notebook and waited for the lecture to start.

'Exciting', he said to Mila.

Dr Kemp arrived, sat at the front, and greeted the students as the late arrivals were still drifting in.

'I hope you found Fanon's book stimulating'. 'It is quite complex, and we will return to this in our last session. For today I want us to question what the purpose of education might be, and if such a purpose, or purposes, could be global. Fanon could help you here of course'.

'Well, that's typical', Alan whispered to Mila, 'We're left to work it out ourselves again!'

References

Abram, D. (1997). *The Spell of the Sensuous*. New York: Vintage Books.

Bainbridge, A., and Del Negro, G. (2020). An Ecology of Transformative Learning: A Shift From the Ego to the Eco. *Journal of Transformative Education*, 18(1), pp. 41–58. https://doi.org/10.1177/1541344619864670

Bainbridge, A. (2019). Education then and now: making the case for *ecol-agogy*, *Pedagogy, Culture & Society*, 27(3), pp. 4232–440 https://doi.org/10.1080/14681 366.2018.1517130

Barnett, R. (2018). *The Ecological University: A Feasible Utopia*. Oxon: Routledge.

Biesta, G. J. J. (2013). *The Beautiful Risk of Education*. Boulder: Paradigm Publishers.

Ingold, T. (1993). 'The Temporality of the Landscape', *World Archaeology*, 25(2), pp. 152–174.

Ingold, T. (2023). On not knowing and paying attention: How to walk in a possible world. *Irish Journal of Sociology*, 31(1), pp. 20–36. https://doi.org/10.1177/079160 35221088546

Moscardini, A.O., Strachan, R. and Vlasova, T. (2020). The role of universities in modern society, *Studies in Higher Education*, doi: 10.1080/03075079.2020.1807493 (accessed 16 December 2020).

Rowlandson, W. (2020). *Living labs, mycelial networks and wild epistemologies. Reflections on emergent pedagogies in Higher Education in a time of crisis*. Paper presented at the Sustainability in Higher Education Conference 'The fierce urgency of now: Navigating paradoxes in sustainability education'. 20–21 May, Canterbury Christ Church University, Canterbury.

Skea, C. and Fulford, A. (2021). Releasing Education into the Wild: An Education in, and of, the Outdoors. *Ethics and Education*, 16(1), pp. 74–90, doi: 10.1080/ 17449642.2020.1822612

Stein, S. (2019). The Ethical and Ecological Limits of Sustainability: A Decolonial Approach to Climate Change in Higher Education. *Australian Journal of Environmental Education*, 35(3), pp. 198–212. doi:10.1017/aee.2019.17.

Sterling S (2021). Concern, Conception, and Consequence: Re-thinking the Paradigm of Higher Education in Dangerous Times. *Front. Sustain.* 2:743806. doi: 10.3389/ frsus.2021.743806

Wallin, P. (2022). Universities are Dead Long Live Higher Education. *Postdigit Sci Educ*. https://doi.org/10.1007/s42438-022-00376-3

12 Researching "Education for Sustainability"

Undergraduate trainees join the dots

Alan Pagden

Introduction

Half a century ago, Illich (1970) presented his vision of a completely transformed society in which the educational relationship between people and their environment is reimagined in the wake of a radical proposal – the abolition of mass compulsory schooling. In arguing now for a transformation of society on a similar scale, the retention of schooling cannot be taken for granted. Indeed, schooling in its current form – built to serve the same ideologically driven agenda that has brought humanity to the brink of catastrophe – is more a part of the "problem" than it is any sort of solution (Sterling, 2019). Here, however, I argue for the retention and radical transformation of schools in parallel with a transformation of society, justified in terms of equity but necessary anyway if humanity is to avert the imminent effects of environmental breakdown and avoid "societal collapse" (Bendall, 2018). A central theme of the chapter is the challenges associated with teacher agency and professionalism in relation to this "transformation". A key to understanding the essence of this problem, I argue, lies in Biesta's (2022) concept of "objectification", a process through which children are reduced to "objects" within systems of "education" focussed narrowly on extraneous goals; in the present context, this means serving the interests of global capital – the "global education measurement system" (Biesta, 2022). In making the case for schooling, it is necessary therefore to recognise the ever-present danger of "objectification", and to develop a conception of teacher professionalism that guards against it.

This chapter is about how best to prepare young people for a career in primary teaching – a career that may extend into the future by more than forty years. In it I argue for a university-based "initial teacher education" since it is only via the "university" component of the Initial Teacher Education (ITE) curriculum that students might acquire the kind of critical awareness they need if they are to act as agents of change in these critical times. To begin with, however, I outline a case for "the transition primary school" – a vehicle and possible end point of radical transformation, since it is in relation to this that the goals of a legitimate initial teacher education (ITE) should be set. This, as will become clear, is a strong claim, but it rests on the best evidence that is available

DOI: 10.4324/9781003286516-18

with respect to the predicament human society is currently in (McGuire, 2022, IPCC, 2022, WWF, 2022, IPBES, 2019) – the "problem" referred to above. Next, I consider the possibility of "a legitimate initial teacher education" within the higher education sector in its current marketized form; I conclude that opportunities still exist on the margins of an essentially broken system. Finally, I describe my experience within this marginal space of supporting a minority of trainee primary teachers who select "Education for Sustainability" as an area for their final year research projects; I provide examples to illustrate the point that, even during times of crisis, there is always positive work that can be done.

The "transition primary school"

Echoing ideas that lie behind the transition towns movement (see Hopkins, 2023), the transition school is one that takes seriously its role in supporting the transitioning of society to a more "sustainable" world. This signals a grassroots agenda for action on the part of stakeholders – teachers, parents, community groups – who, in the absence of an appropriate and timely response from policy makers, do what they can to enact necessary change. Working within – but also to some extent "resisting" (Biesta, 2015) – the constraints of the neo-liberal policy agenda (Hall and Pulsford, 2019) those schools that prioritise "sustainability" in their policies, curricula, and community resilience-building are described here as "transition schools". The concept of the "transition school", however, also serves as an invitation to consider what primary schools might look like in a "sustainable" society of the future, impossible to describe in detail, but one nonetheless that would be founded on the principles of "eco-socialism" (Fraser, 2021), or equivalent (see for example, Hickel, 2021). Although it is vital to maintain a sharp focus on the urgency of the present – as Biesta (2022) points out, children cannot be expected to wait for the "ideal" education of a hoped-for future – current practices and strategies of resistance should be informed by these principles. Hence, whilst I begin this section by exploring the characteristics of schools that fit the first definition – those that currently prioritise "sustainability" – I then move on, via three key authors (Fraser, 2021, Sandal, 2020 and Biesta, 2022) to highlight an additional set of principles that would necessarily underpin the purpose and practice of primary education within an eco-socialist society, even though under current circumstances alignment with these principles is especially challenging.

One useful model for thinking about the characteristics of existing "transition schools" can be found in Assadourian's (2017) version of "Earth Education". This is summarised in the form of two diagrams. First, Assadourian's (2017) six "Earth Ed" principles are represented as a pyramid, loosely based on Maslow's hierarchy, with "earth dependency" at its base. This aligns strongly with my preferred definition of "sustainable development" (Griggs et al. 2013, 206) which, in a similar vein, emphasises the primacy of the biosphere, "earth's life support system". Assadourian's (2017) foundational premise of "earth dependency" asserts the need for children to develop an understanding of our complete dependence on the biosphere and our current need to restore and protect it, something that generations of adults in the Global North have forgotten. Second, alongside the pyramid, Assadourian (2017) positions "Earth Ed" at the centre of a Venn diagram, where it brings together two overlapping domains, "Education for Sustainability" (EfS) and "Education for Resilience" (EfR). Whilst the purpose of the former is to produce "sustainability champions", the purpose of the latter is to "make students more resilient to the changes that are locked into their future" (Assadourian, 2017, 7). A key strength of the Earth Ed model lies in its attempt to bring these two domains together.

Assadourian's (2017) Venn diagram, furthermore, has intuitive appeal, since it suggests the integration of two very different kinds of activity that typically occur in primary schools; these are, on the one hand, the teaching of relatively abstract concepts such as "the carbon cycle" or "biodiversity" (EfS) – arguably requiring classroom-based "seat work" of one kind or another – and, on the other hand, practical hands-on activities that engage children directly with the non-human living world (EfR) including Forest School and food education projects. The overarching goal of Earth Ed is to achieve a seamless integration of these domains from the child's perspective, so they can bring understandings from the classroom – for example, about the build-up of GHGs in the atmosphere – to their direct experience of the bio-physical world – for example, in Forest School where they engage directly with fire. Unfortunately, integration remains a distant goal in some schools where, despite their best intentions, Forest School is introduced as an add-on, detached in experiential terms, from the content and priorities of the classroom-based curriculum (Kemp and Pagden, 2019). This illustrates the need for a whole curriculum approach based on a clearly articulated sense of educational purpose (Biesta, 2015), arguably something that is difficult to achieve within current policy frameworks.

Whilst there are several aspects of the Earth Ed model that are worth exploring in more detail, three are of particular relevance to my argument. Two of these, "systems thinking" and "critical consciousness", are core components of EfS, as described above, while the third, "life skills" more obviously falls within the domain of EfR. The case for developing "systems thinking" in children (Booth Sweeny, 2017), in my view, provides the strongest justification for what, in its various guises, is commonly referred to as "cross-curricular teaching and learning". More important, however, in the primary years, a focus on "systems thinking" should be seen to lay the foundations for an understanding of how "socio-ecological systems" work at different scales within the Earth system, enhancing and/or undermining its "resilience" (Folke et al., 2010). Whilst the concept of "resilience" is contentious and open to misuse – for example, through policies that encourage "self-objectification" in children (see below) – from a systems perspective, it provides a lens through which to view the school as a socio-ecological system more or less resilient within its surrounding community; indeed, "resilience" is deployed in this sense within the transition towns movement (Hopkins, 2009, 12, cited in Dixon, 2022, 174) . Furthermore, "systems thinking" is not only about understanding how complex relations between multiple entities determine how the world works: it includes an understanding of how the world can be changed, for the better.

Developing "critical consciousness" in pupils is a key goal of Earth Ed and it is important to consider how this might be achieved. Teachers who are both knowledgeable and deeply aware of the "polycrisis" that we face, will understandably want to convince their pupils of its urgency; they might also seek to encourage activism of one kind or another. For Assadourian (2017, p. 15), "critical consciousness" involves being able to "perceive the hypocrisies" in current systems which "need to be corrected … to help create a just and sustainable society". Clearly these are insights and convictions that children need to arrive at independently if at all, but schools should provide the conditions within which this can happen. Most obviously there is a need for sound citizenship education – education for democracy – where, through debating controversial issues (see Crick, 1998, p. 56–60), children are empowered, both individually and collectively, to develop their own perspectives on the issues discussed. The teacher's role in this context is to provide "open and safe spaces for dialogue and enquiry" (Andreotti, 2006), something that can easily be achieved within the bounds of their legal obligation in the UK to maintain "political impartiality" (see DfE, 2022). Indeed, as Biesta (2022) would argue, a genuine educational experience is one where children – as subjects – are free to decide for themselves what to do with what is put in front of them. Children's freedom, to be clear, is not an unqualified freedom; it is constrained by the fact of living in an ecologically vulnerable world with other people (Biesta, 2022). The task of education is to confront the child with their freedom (and responsibility) in this regard. In addition to safe spaces to discuss contentious societal issues, schools need to establish, within their everyday practices, genuinely – i.e., nontokenistic – democratic decision-making processes that all stake holders participate in.

Whilst the term "life skills" is used by Assadourian (2017) in a relatively loose way, here – taking food education as a pivotal case – I focus on the acquisition of skills that are essentially of a practical nature. A comprehensive food education programme should lie at the heart of the transition school curriculum, and it is not difficult to find schools that exemplify this (see for example Washingborough, 2023, and Charlton Manor, 2023). Informed by an understanding of the present food systems crisis – food security, environmental breakdown, and poor diets – the food curriculum in primary schools should consist, in the main, of practical multisensory experiences in the garden, the kitchen and the dining room. Furthermore, since we eat every day, it should permeate virtually every aspect of school life. Although space limitations prevent me from exploring food education practice in any detail, here I highlight two strategies that exemplar schools typically adopt, both of which illustrate something important about the transition school. First, when addressing sustainability and nutrition issues schools take full responsibility for the preparation of meals, their procurement decisions invariably favour local producers, and this leads to a strengthening of community bonds. Procurement more generally should be seen as an opportunity for schools to build community cohesion and resilience (Sustainable Food Trust, 2022). Second, it is not uncommon for schools to spend a portion of their limited budgets on employing non-teaching adults in roles to support their food programmes – most obviously, gardeners and chefs. If, as I argue, children benefit from being exposed to adults in a range of non-teaching roles, a goal for the transition school might be to increase this range by diversifying its activities; this idea is explored further below.

Whilst the Earth Ed model provides a clear aspirational vision for the transition school – other possibilities include "Nature's Principles of Harmony" (Harmony Project, 2023) the UN's SDGs (UNESCO, 2023) and the Earth Charter (2023) – my account, so far, is neither comprehensive nor concrete. Dixon (2022), on the other hand, provides a more complete exposition of the "transition school" and although he does not adopt the term, as I have done here, he does draw a parallel with transition towns. Through his five Cs – Captaincy, Curriculum, Campus, Community and Connections – he explains how, under his leadership, two very different primary schools made progress across all five Cs towards key "sustainability" goals. Dixon's (2022) is a powerful counter argument to anyone who maintains that sustainability goals are impossible to achieve under the current neoliberal policy framework; he shows how schools under the right leadership can work within and take advantage of policy and, at the same time, exploit the positive opportunities that their unique circumstances offer. Each of his schools, occupying a unique location – geographically, historically, and socially – affords its own range of opportunities for building community resilience. This lends weight to the argument that, contra the recent, COVID related, enthusiasm for online alternatives, "physical, local schools" have a vital role to play in building "intergenerational solidarity, respect for diversity and democratic capability" (Facer, 2011, p. 28). Dixon understands the demands and constraints of neoliberal policy – he even suggests that school leaders should educate themselves about its origins (Dixon, 2022, p. 16) – but ultimately, he is successful only because he is compliant and, in some instances – for example, in the celebration of the schools' success with high stakes testing – he may not be fully aware of where this occurs. Hence, the need to spell out the extent to which schools could be transformed in parallel with the systemic transformation of society away from neoliberal capitalism.

The "transition primary school" of the future

I turn now to present key ideas from three authors – ideas which taken together sign-post the truly radical transformation that schools will need to undergo as we transition to an eco-socialist future. The implication here is that this transition *will* take place since the greening of capitalism is now impossible (Calverley and Anderson, 2022; Hickel, 2021) and the alternative, in one form or another, is societal collapse. First, I draw on Fraser (2021), whose analysis of capitalist society implies a major rethink of how work is organised and rewarded; this has implications for the kinds of responsibilities that adults will take on in primary schools in the future. Next, I turn to Sandal (2020) whose critique of meritocracy challenges certain fundamental beliefs about how children should spend their time in school and about who they should spend that time with. Finally, I look more closely at Biesta's (2022) argument for a "world-centred education", focussing primarily on the nature of the teacher-pupil relationship, without which there would be no "education". To reiterate, although my intention in this section is to illustrate possibilities for schooling in a non-capitalist future, the ideas presented are intended primarily to inform teachers' and student teachers' thinking, in the present.

Fraser (2021) unequivocally, and in my view correctly, identifies "capitalist society" as the root cause of the "general crisis" that humanity finds itself in. Through a historically grounded analysis of how relations and contradictions within capitalist society work, she explains the necessity of anti-capitalist struggle as the only hope for the future survival of our species. There are two important steps in Fraser's (2021) argument that point to key principles for the transition school. The first springs from her call for a "trans-environmental" approach to activism and the second arises in the context of her conceptualisation of "capitalist society". Although the main focus here is on the second "step", I shall briefly explain the relevance of the first. Fraser's (2021) assertion that anti-capitalist struggle must be "trans-environmental", including the voices of all who fight for justice under the current system, mirrors the entanglements that exist between the environmental and other crises. For the transition school the implications are twofold: first, the curriculum should be trans-environmental and interdisciplinary; and second, the school should be genuinely inclusive. Since these principles are well understood by pioneering school leaders (see Dixon, 2022) I will not explore these points in any depth. However, in anticipation of the chapter's final section, I do note that important connections between different aspects of the "general crisis" or "polycrisis", that Fraser's (2021) analysis brings to the fore – for example, the link between our colonial legacy and global heating – are not always obvious to trainee teachers.

"Capitalist society", according to Fraser (2021), is a term that encompasses more than "capitalism", the economic system per se; it includes the specific set of social and political conditions without which that economic system could not exist. These conditions include relations based on the legitimisation of forms of exploitation and expropriation which take as given the premise that certain goods and services are free, including the biosphere and social reproductive labour. Whist protection of the first of these "free" goods is the target of environmental activism and an obvious focus for the curriculum of the transition school, the question as to how social reproductive labour should be organised from an eco-socialist perspective requires a more nuanced response. In a society based on non-exploitative forms of labour, social reproductive work, including work with young children, would be highly valued whether this work took place in the home or the school. The question as to whether all valued work should be rewarded as paid employment adds a layer of complexity that I will not explore in detail here apart from acknowledging proposals, consistent with eco-socialism, for either a "universal basic income" (UBI) or guaranteed work (Hickel, 2021) either of which would generate new possibilities for the employment of adults in primary schools.

This signals a need to revisit assumptions about the fundamental purpose of schooling. As the recent COVID pandemic demonstrated, the function of schools in society is not – and probably never was – confined to the narrow task of "delivering" a curriculum; schools also exist to ensure that young children are cared for, a task that contributes significantly to society's overall resilience in the face of challenges, such as COVID. This reinforces the point made above that in an eco-socialist future caring – a dimension of social reproduction – would receive the recognition that it deserves. One consequence, in schools of the future, would be an expansion of adult roles that have a caring component however these roles might be defined. In addition to this, as suggested above in the discussion about food education, there would be an expansion of roles in other areas, many of which would arise in the context of additional strategies for enhancing community cohesion and resilience. Imagine, for example, the potential of employment opportunities within initiatives to enhance biodiversity in specific localities, where the school, at the heart of its community, could play a pivotal role. How children might "benefit" from working alongside adults who, in a formal sense, might not be teachers is an important question which I look at briefly below.

As Sandal (2020) explains, the "tyranny of merit" dovetails quite well with the neoliberal agenda for education whereby schools and other "providers" compete within a marketplace for "customers" – children via their parents, HE students on their own account – who themselves compete in a race for the rewards of capitalism – money and status – which are divvied out, ostensibly in a just way, based on merit. Sandal (2020) explains the downsides of being a winner in the hyper-competitive race for success that the offspring of the elite are forced to participate in, but the most pernicious aspect of this system is its effects on the "losers", who come to believe that they are ultimately responsible for their own failure. This effect is then inadvertently exacerbated in schools by teachers who, through efforts to encourage children to act responsibly, adopt practices – for example, those based on ideas about "mastery", "growth mind set" and "self-regulation" – that serve the dominant policy agenda, which in this context is to create "little neoliberals" (Bradbury, 2019) – young children who, as entrepreneurs of their own destiny, ultimately have no one else to blame but themselves for their relative success or failure. This process, described by Biesta (2022) as a form of "self-objectification", shows the magnitude of the dilemmas that teachers currently face and the degree of critical insight that is needed going forward.

As Sandal (2020) convincingly argues, even if it were possible to create a genuinely level playing field, rewarding some people more than others cannot be justified on the presumption that a combination of talent and hard work implies a higher level of deservedness. Clearly, however, in a complex society the need to reward some people on the grounds of merit may persist – for example, in areas of work that require specialist knowledge and skills – but this should be done in the interests of society at large, not because these individuals are regarded as more worthy than their fellow citizens. Sandal (2020) envisages a society that, in the interests of "the common good", is far more equal and, as he argues, to achieve this the principle of "equality of opportunity" is insufficient. Citing the philosopher Tawney (1964) he argues for "equality of condition" as the basis for social life, a principle that is fully consistent with the eco-socialist agenda. On this account, the transition primary school should reject the implicit societal assumption that there is something inherently good about meritocracy and it should push back against the impetus to place children in a perpetual race to outcompete their peers.

In transition schooling moving forward, changes that need to be made with a view to dismantling the structures that underpin "meritocratic hubris" (Sandal, 2020), include the following. To begin with private schools as found in England would be abolished as would high stakes testing and draconian inspections. There would need to be an overall shift away from the language of "ability" with its associated practices (see Hart et al., 2004). Furthermore, and more fundamentally, the taken-for-granted and currently predominant way of structuring schools through age-based cohorts that determine the grounds for judging educational progress – that is, through comparison between same-age peers – would be reconfigured; although it is difficult to say exactly how this might be done, the expansion of the adult work force (see above), would enhance the possibilities. These changes would bring the school more in line with Biesta's (2015, 2022) depiction of it, as a protected space between the home and the street where children have the opportunity to "practice being grown up". A key dimension of this "space", emphasised by Biesta (2022) in his recent writing, is that of "suspension", the idea in essence that children need *time* to practice, including time to get things wrong. This has implications for how children's time is managed, including with respect to who they spend their time with; hence, the current formula of sequenced lessons, delivered to groups of thirty same-age peers, will likely be a rare event in the transition school of the future.

Biesta's (2015) model of educational purpose, mentioned above, cuts through the age-old, often paralysing, debates in education that pivot on the false binary of progressive versus traditional practice. His three domains of purpose – qualification, socialisation and subjectification – emerge from his critique of the still prevalent discourse of "learning" (Biesta, 2015). In his most recent writing, he has both elevated the importance of the third domain – subjectification – and clarified misconceptions that have been associated with it (Biesta, 2022). Education, for Biesta (2022), is an existential matter, the core question being one of how the child can "exist as subject *in* and *with* the world". In other words, it is a question of how the child can exist in a "grown-up" or "non-ego-logical" way, on a vulnerable planet with other people. Subjectification – exiting as subject – involves the exercise of freedom, albeit a qualified freedom that is integral to any genuinely educational teacher-pupil relation as explained above in the discussion about education for democracy. The teacher's task, in this context, is to bring the world to the child who, in presenting themselves as one who can be taught, may be surprised, or moved by the encounter; they may, on the other hand, simply turn the other way.

For the transition school it is important to note that there are different ways in which the teacher can fulfil this fundamental task; in the example of "education for democracy" it is by providing "open and safe spaces for dialogue". Additionally, however, the teacher, especially the primary school teacher, must provide the conditions for children to be taught directly by the world, acting in its own right; this includes the inanimate world of tools and materials and the non-human living world of animals and plants (Biesta, 2022, p. 99). As the Earth Ed model and the food education example discussed above make clear, opportunities of this kind should form a significant part of the transition school curriculum. A matter that requires further clarification, however, is the role that adults would play in relation to this component of the curriculum – learning directly from the world.

Although the proposal for an expanded work force suggests multiple adult roles – chefs, gardeners, librarians, repair workers – of the kind that children would learn from through "observation and pitching in" (see Rogoff, 2003 on the process of "guided participation"), I argue that a case must still be made for the professional, "qualified" teacher. The teacher as "professional" should exhibit a minimum level of relevant expertise but as mentioned above, in relation to the process of "self-objectification", they should also demonstrate a deep critical awareness of the contexts in which they work. Ultimately their task will be to sustain the school as a place where children have genuine opportunities to "practice being grown up" (Biesta's, 2015), a challenge that requires at least two things: first, they will need to have the capacity to harness the skills of their non-teaching colleagues through a deep understanding of "guided participation" and sound leadership; and, second, since the power in adult-child relations is asymmetrically distributed and always will be, they will need to be proactive in guarding against the possible re-emergence of "objectification" in a system that will continuously evolve. Working towards a redefinition of professionalism is a task that teachers will need to undertake collectively, which is an enormous challenge, as the work of Moore and Clark (2016) demonstrates. Furthermore, as this and the previous section are intended to illustrate, teachers need a clear direction of travel.

University-based Initial Teacher Education (ITE) in 2023

The case for a university-based initial teacher education, in my view, can be made on the grounds that it combines "training" with the critical insight that is a vital component of the emergent "professionalism" described above. One way of thinking about this added dimension is in terms of a traditional "liberal education" (Hussey and Smith, 2010, p. ix) which "can be achieved in any discipline ... so long as the primary focus is on the critical evaluation of the content and not just the content itself". By positioning students as customers, the remit of the university has shifted from education (and research) to one of "servicing needs". In defence of "liberal education", Biesta (2022, p. 70) describes this as "giving customers what they want" rather than doing the truly educational thing, which is to give "students what they didn't ask for, first and foremost because they didn't even know they could ask for it". This, of course, echoes the description above of the teacher's task which is to bring the world to the child who should be free to do with it what they will. As customer the student is less inclined to "present themselves as one who can be taught" and more inclined to focus on the commodity that they have purchased – the degree and the degree classification that they hope to receive. This, together with unsustainable workloads, has undermined the lecturer's capacity to confront the student with their freedom, which is another way of describing the essence of a "liberal education".

Another effect of marketization is that some universities appear to have lost their way by abandoning their responsibility to act as gatekeepers to the professions. In a complex society, as Sandal (2020) explains, some degree of selection by merit is both necessary and inevitable. Hence, in a society founded on the principle of "equality of condition" everyone would have the right to access good quality education throughout their lifespan – including a diverse range of post-compulsory offers – but this, to be clear, does not imply a universal right to "higher education", or the automatic right to practice in any profession. Ignoring this definitional premise, that universities are non-compulsory and selective, can lead to confused thinking whereby the purpose of the university is conflated with that of the comprehensive school, to provide a general education for all; see the recent – albeit well intentioned – report by Blake, Cooper and Jackson (2022), which is a clear exemplification of this phenomenon. For those of us who work in university-based ITE in England the consequence is that our role as "gatekeepers to the profession" has been diminished and, in effect, outsourced to the Department for Education (DfE).

Compounding the effects of marketization on the student-tutor relation ITE lecturers, working in HE in England, endure additional pressures to induct students into the "culture of compliance" – identified by Alexander (2010) in his landmark report on primary education – that is still prevalent in schools. It is an irony of neoliberalism's quasi-market in education that to function it requires a highly centralised and tightly regulated bureaucracy. In England qualified teacher status (QTS) can be obtained in a variety of ways, only one of which is via a course of study in an HE institution; the qualification to teach is accredited by the DfE against centrally defined criteria and "standards" are strictly policed by the Office for Standards in Education (Ofsted) the same inspection body that inspects schools. Recently the department imposed an official curriculum on all initial teacher training (ITT) providers (DfE, 2019) – notice the shift from education to training – although it insists that this is not a "curriculum" and that providers are free to deliver additional material if they want. The regulations, however, require a minimum number of days for students to spend in school where, one way or another, they are judged according to how well they perform in the classroom. In this regard, "performance", in no small part, means compliance with the hegemonic agenda outlined at the start of this chapter, which is to "objectify" children in the interests of global capital. Hence, my conclusion that the system is "broken".

ITE students conducting small-scale research in schools

In this final section I briefly describe a "marginal space" wherein a small, self-selected group of trainee primary teachers encounter the challenges of transition schooling via a research process which does, in certain key respects, match up to the tenants of a genuine "liberal education". This opportunity occurs within a final year undergraduate research module framed by a discourse of professionalism, research-informed practice and continued professional development; a central aim of the module is to inspire students to believe in the value of researching their own practice. From a teaching perspective, there are three aspects of this research process where the goal – albeit partially implicit – is to expand the students' understandings: first, in conducting the research students become more attuned to the perspectives of their pupils; second, by paying close attention to their specific "research context", students deepen their understanding of the multidimensional nature of transition schooling; and, finally, by "joining the dots" between their small-scale project and the global polycrisis, students' belief in their agency as teachers is strengthened. I turn now to illustrate each of these "goals" through three concrete examples.

By adopting a small-scale, qualitative, participant-centred research methodology students expand their understanding of educational research possibilities whilst refining their listening/observation skills. In my experience students often begin by thinking that "research" must involve some form of controlled experiment or survey; typically, their initial research questions are formulated in terms of "effect" or "impact". Through a process of negotiation – including input on small-scale qualitative research methodology – they refine their questions and, in most cases, settle on something that requires the elicitation of children's perspectives/understandings. They then devise age-appropriate elicitation tasks – using for example drawing, photography, guided tours, or map-making – around which meaningful conversations can ensue. Tasks are, for the most part, similar to the kinds of classroom activities that children routinely participate in but, with their researcher's hat on, student teachers are able to pay closer than routine attention to the meanings generated by the children. One student whose question arose in the context of an existing classroom project on "recycling" with four- and five-year-old children, began by inviting children to explore a collection of items of "rubbish". She was surprised to find that some children initially lacked the vocabulary to describe what certain objects are made from (wood, metals, plastics); this led her to conclude that the project should be redesigned to incorporate more preliminary tactile activities using a range of objects/materials.

By focusing in detail on the unique circumstances of their school – their "research context" – students gain insight into the multi-dimensional nature of transitional schooling and establish a firm basis for the interpretation of "data". This is exemplified through projects that look at aspects of food education, including initiatives focusing on growing, food preparation, or the recycling of waste. Once students are introduced to the multifaceted nature of the challenges via exemplar schools (see above) they realise the unique situation that their individual schools are in and the implications that this has for their choice of project. One student, working with seven- and eight-year-old children, was initially disheartened when she found that food education was not considered a priority in her school. She was, however, able to devise elicitation tasks to explore children's understandings of different food items and their origins. Through her analysis of the school context, she identified key factors – cultural difference and poverty – that had a bearing on how individual children experience food. These insights helped her to frame her project and interpret her findings. As well as demonstrating the importance of listening to children's unique stories about food, she also realised the value of discussions that bring diverse perspectives together. In a school that pays little attention to food education or the issues that it is intended to address, listening to children is a good place to start.

Finally, by making connections – "joining the dots" – between their small-scale projects and the urgency of our global predicament – the "brief and rapidly closing window of opportunity [that we have] to secure a liveable and sustainable future for all" (IPCC, 2022, p. 35) – students strengthen their sense of agency. This is achieved (to the extent that it is achieved) via a two-way process. To begin with, most students arrive with a specific interest in something to do with "sustainability", motivated by the idea that this could be a worthwhile focus or context for their teaching. They are then introduced to material that is intended to heighten their awareness of our global predicament and, in all cases, through no fault of their own, this is an eye-opener. One student, interested in biodiversity loss, decided to look at representations of the non-human living world in children's popular culture. She found a set of picture books aimed at counteracting the effects of stereotypical images by portraying animals in authentic ways (Walker, 2023). Working with eight- and nine-year-old children, she organised a series of activities to facilitate discussion about how these representations work in contrast to the many other images that the children are familiar with. Even though this is clearly a worthwhile endeavour, it addresses Assadourian's (2017) first principle – the student expressed uncertainty about the relevance of the topic, given its position within the "literacy" curriculum. After completing her written report, the same student commented on how, as a consequence of undertaking her research, she now appreciated the significance of biodiversity loss, and this had reinforced her commitment to teaching children about it.

Conclusion

This chapter began with the assertion that the goals of initial teacher education in the primary sector – and, by extension, primary teaching in general – should be set within the parameters of a case for the "transition school", one version of which is outlined above. This was followed by an argument for a university-based "liberal initial teacher education" as a necessary foundation for the "new professionalism" that is required to steer schooling away from the constraints of neoliberalism and towards an eco-socialist future. In the final section I described a research module on "education for sustainability" that is offered to undergraduate primary teacher trainees. From my perspective this module has the potential to offer an element of "liberal education" within an ITE programme that is, nonetheless, dominated by an externally determined drive for compliance. A key message that runs through the examples of student research projects, is that wherever teachers work there is always something that they can do to shift their practice and the practice of their schools in the right direction. Clearly this requires us – here I include myself as a university lecturer – to have an understanding of our predicament and a set of principles to provide direction. More than this, however, it requires us to sit comfortably with the fact that our students will not always present themselves as "one[s] who can be taught"; this is the "beautiful risk of education" (Biesta, 2022), an unavoidable risk, in an era that demands our respect for freedom.

References

Alexander, R. (ed.) (2010) *Children, Their World, Their Education.* London and New York: Routledge.

Andreotti, V. (2006) *Critical Literacy in Global Citizenship Education.* Open Spaces for Dialogue and Enquiry Available at: www.academia.edu/194048/Critical_Literacy_in_Global_Citizenship_Education_2006_ (Accessed: 13 February 2023).

Assadourian, E. (ed.) (2017) *EarthEd: Rethinking education on a changing planet.* Washington, DC: Worldwatch Institute.

Bendell, J. (2018) *Deep Adaptation: A Map for Navigating Climate Tragedy* IFLAS Occasional Paper 2 Available at: www.cumbria.ac.uk/research/centres/iflas/ (Accessed: 13 February 2023).

Biesta, G. (2015) "The duty to resist: Redefining the basics for today's schools". *ROSE,* vol 6 Special Issue pp. 1–11.

Biesta, G. (2022) *World-Centred Education: a view from the present.* New York and London: Routledge.

Blake, S. Cooper, G. and Jackson, A. (2022) *Building Belonging in Higher Education: recommendations for developing an integrated institutional approach.* Pearson Wonkie.

Booth Sweeny, L. (2017) "All systems go! developing a generation of 'systems smart' kids", in Assadourian, E. (ed.) *EarthEd: Rethinking education on a changing planet.* Washington, DC: Worldwatch Institute.

Bradbury, A. (2019) "Making Little Neo-liberals: the production of ideal child/learner subjectivities in primary school through choice, self-improvement and 'growth mindsets'". *Power and Education* 11(3): 309–326. https://doi.org/10.1177/175774381 8816336

Calverley, D. and Anderson, K. (2022) *Phaseout pathways for fossil fuel production within Paris-compliant carbon budgets.* Tyndall Centre, University of Manchester.

Charlton Manor Primary (2023) School Website Available at: https://charltonmanor primary.co.uk/ (Accessed: 13 February 2023).

Crick, B. (1998) *Education for Citizenship and the Teaching of Democracy in Schools: final report of the Advisory Group on Citizenship.* Qualifications and Curriculum Authority.

DfE (2019) *ITT Core Content Framework.* Department for Education.

DfE (2021) *Teacher Standards.* Department for Education.

DfE (2022) *Political Impartiality in Schools Department for Education.* London: HMSO. Available at: www.gov.uk/government/publications/political-impartiality-in-schools/ political-impartiality-in-schools accessed 24 June 2023.

Dixon, D. (2022) *Leadership for Sustainability: saving the planet one school at a time.* Carmarthen: Independent Thinking Press.

Earth Charter (2023) Available at: https://earthcharter.org/ (Accessed 13 February 2023).

Facer, K. (2011) *Learning Futures: education, technology and social change.* London and Routledge.

Folke, C., Carpenter, S.R., Walker B., Scheffer M., Chapin T., and Rockström, J. (2010) "Resilience Thinking: Integrating Resilience, Adaptability and Transformability", *Ecology and Society* 15(4): 20. [online] URL: www.ecologyandsociety.org/vol15/iss4/ art20/ (Accessed: 13 February 2023).

Fraser, N. (2021) "Climates of Capital: for a trans-environmental eco-socialism", *New Left Review*, 127 Jan/Feb 2021, pp. 94–127.

Griggs, D., Stafford-Smith, M., Gaffney, O., Rockström, J., Öhman, M.C., Shyamsundar, P., Steffen, W. Glaser, G., Kanie, N., and Noble, I. (2013) "Sustainable Development Goals for People and Planet", *Nature* 495(7441): 506–507. https://doi.org/10.1038/ 495305a

Hall, R. and Pulsford, M. (2019) "Neoliberalism and primary education: Impacts of neoliberal policy on the lived experiences of primary school communities", *Power and Education* 11(3): 241–251.

Harmony Project (2023) *Putting Sustainability and Nature at the Heart of Learning Available* at: www.theharmonyproject.org.uk/ (Accessed: 13 February 2023)

Hart, S., Dixon, A., Drummond, M. J. and McIntyre, D. (2004) *Learning without Limits.* Maidenhead: Open University Press.

Hickel, J. (2021) *Less is More: how degrowth will save the world.* London: Penguin.

Hopkins, R. (2009) *The Transition Handbook* Cambridge: Green Books.

Hopkins, R. (2023) Rob Hopkins: Imagination Taking Power. Available at: www.rob hopkins.net/ (Accessed: 13 February 2023).

Hussey, T. and Smith, P. (2010) *The Trouble with Higher Education: a critical examination of our universities,* London: Routledge.

Illich, I. (1970) *De-schooling Society*. London: Penguin.

IPBES (2019) *Summary for policymakers of the global assessment report on biodiversity and ecosystem services of the Intergovernmental Science-Policy Platform on Biodiversity and Ecosystem Services* IPBES.

IPCC (2022) *Climate Change 2022: Impacts, Adaptability and Vulnerability: Summary for Policymakers*. IPCC.

Kemp, N. and Pagden, A. (2019) The Place of Forest School within English Primary Schools: senior leaders perspectives. *Education 3-13*, 47:4. https://doi.org/10.1080/03004279.2018.1499791

McGuire, B. (2022) *Hothouse Earth: an inhabitant's guide*. London: Icon Books.

Moore, A. and Clark, M. (2016) "'Cruel Optimism': teacher attachment to professionalism in an era of performativity". *Journal of Education Policy*, 31:5 pp. 666–677 https://doi.org/10.1080/02680939.2016.1160293

Rogoff, B. (2003) *The Cultural Nature of Human Development*. Oxford and New York: Oxford University Press.

Sandal, M. (2020) *The Tyranny of Merit: What's Become of the Common Good?* London and New York: Allen Lane.

Sterling, S. (2019) "Planetary primacy and the necessity of Positive Dis-Illusion", *Sustainability*. Vol. 12(2) pp. 60–66.

Sustainable Food Trust (2022) *Feeding Britain from the ground up*. Sustainable Food Trust. Available at https://icdasustainability.org/report/feeding-britian/#:~:text=Feeding%20Britain%20from%20the%20Ground%20Up%20%E2%80%93%20for%20climate%2C%20nature%20and,farming%20based%20on%20biological%20principles. Accessed 24 August 2023.

Tawney, R.H. (1964) *Equality*. London: Unwin.

UNESCO (2023) *Sustainable Development Goals*. https://en.unesco.org/sustainabledevelopmentgoals (Accessed: 13 February 2023).

Walker (2023) Available at: www.walker.co.uk/UserFiles/file/Nature%20Storybooks/4944%20Nature%20storybooks%20poster%20A2_08%20.pdf (Accessed: 13 February 2023).

Washingborough Academy (2023) School Website Available at: www.washingborough academy.org// (Accessed: 13 February 2023).

WWF (2022) *Living Planet Report 2022: Building a Nature Positive Society*. WWF International.

13 What we must do now

The response(ability) of universities to the global crises

Nicola Kemp

Introduction

Having powerfully presented evidence about the urgency of the climate crisis, Thunberg's (2022), *The Climate Book*, challenges the reader with the question 'what must we do now?' However, as Fraser (2021, p. 95) reminds us, we should avoid treating 'global warming as a trump card that overrides everything else' and rather consider 'all major facets of this crisis, "environmental" and "non-environmental"'. Similarly, physicist Fritjof Capra reminds us that 'the major problems of our times [are] just different facets of one single crisis, which is largely a crisis of perception' (cited in Davidson, 2020, p. 21). In this chapter, I focus on responses from the higher education sector to the contemporary polycrisis and offer a reflective account of the experience of one university.

The scale and pace of the global environmental crisis have led many universities to develop institutional commitments to change in recent years. Commonly this involves being a signatory to the United Nations Sustainable Development Goals (UN SDG) accord and/or declaring a Climate Emergency, invoking three types of 'adjustments' to practice: developing solutions-focused research, embedding Education for Sustainable Development (ESD) in the curriculum, and reducing the ecological footprint of the institution (Stewart, Hurth and Sterling, 2022). However, such responses are essentially reformist and have allowed sustainability to be added on or accommodated within existing structures whilst maintaining the status quo. Authors, including Sterling (2021, p. 3), argue that this approach has rendered sustainability 'safe' and tame, and limited the possibility for universities to develop more meaningful responses. Instead, he suggests, the 'wicked problem' of unsustainability requires first, a questioning of purpose. This questioning needs to be both *retrospective*, acknowledging the involvement of universities in 'allowing this systematic unsustainability to emerge' (Stewart, Hurth and Sterling, 2022, p. 1), and *prospective* in seeking to understand how they should position themselves in relation to their fundamental missions of education, research, and civic contribution – a re-purposing. A recent special edition of the journal *Frontiers in Sustainability* suggests this re-purposing should be 'for Sustainable Human

DOI: 10.4324/9781003286516-19

Progress' contributing to the well-being of human and non-human life. This, the editors suggest, would require a 'radical reimagining' of the university.

This chapter will explore the question of how higher education can be re-purposed 'as if the world mattered'. Drawing upon practical experience, together with insights from research and theory, it considers the challenges and offers a response to the question raised by Stewart, Hurth and Sterling (2022, p. 1) of 'how prudent, meaningful change might be operationalised at scale and pace'.

Institutional paradoxes

Assuming a university has 'transformational intent' (Fazey et al, 2021), the question of how to operationalise this is a complex one. I have argued previously that there are two fundamental contradictions or paradoxes that universities and those who work in them need to consider '(1) How to develop authentic sustainability responses within the context of existing higher education structures and processes (2) How to reconcile the demand for immediate action with the much more gradual processes of education' (Kemp and Scoffham, 2022, p. 6). These are represented as intersecting axes on a diagram (see Figure 1.1). The model was purposely intended for use at different levels – the individual, team, and organisation. In this chapter, the focus is explicitly on the organisational or institutional response, so it is worth briefly outlining the scope of each of the paradoxes at this 'level'.

1. **The resistance/alignment paradox**

   Universities operate within the same dominant neoliberal paradigm that is recognised as contributing to the systemic unsustainability they are seeking to change (Bessant, Robinson and Ormerod, 2015). A key tension for universities is that they need to continue to deliver their essential mission (aligning with the current political economy) whilst changing (resisting it). A recent report by the Sustainable Development Solutions Network (2021) posits that the transition towards sustainability may necessitate both 'alignment' to ensure the continuation of key functions alongside 'resistance'; the suggestion being that they can be complementary responses. The key question at an institutional level may therefore be when to resist and when to align. As Biesta (2022, p. 11) argues, the move towards 'world centred education' requires an understanding of 'whether the particular circumstances are worth adapting to, or whether there is a need to resist and refuse adaptation'.

2. **The fast/slow paradox**

   A further tension for universities is managing the tendency to focus on fast variables that can yield immediate, and observable (and arguably superficial) change rather than the slow variables that can drive transformation – this is the temporal paradox. I have argued previously that the educational process can itself be considered a slow variable that is at odds with the urgency of the environmental crisis (Kemp and Scoffham, 2022). However, at an institutional level, there is a range of other influential slow variables, both within and beyond the direct control of an individual university. Robinson and Laycock Pedersen (2021), for example, detail the challenge of changing academic traditions and organisational culture within universities. 'As well as an organisational culture of the institution as a whole, the different domains of academic activity, education, research, campus, outreach as well as different disciplines, are all marked by differences in culture' (p. 10). They also highlight the influence of 'external' variables such as international and national policies and regulations that can either support or limit the repurposing process.

Navigating paradoxes through a second operating system (SOS)

One approach to navigating these institutional paradoxes proposed by the Sustainable Development Solutions Network (2021, p. 39) is to adopt a 'second operating system'. This adapts the concept of the dual operating system developed by Kotter (2012) as a way of supporting business change to the context of higher education. It is worth presenting the concept as articulated in the report as it starts by acknowledging the critical point that universities need to ensure continuity of their mission (education, research and civic contribution) throughout the period of transformation.

> Since Universities need to continue to deliver their essential mission, implementing organizational reform at scale must not come at the expense of a delay or halt to their day-to-day activities ... For this reason, one approach could be to develop a kind of 'second operating system'. This ... would be focused solely on designing the appropriate transformation that could complement the existing governance system of the university ... the second operating system can work as an 'agile, network-like structure and a very different set of processes'.

Features of this SOS include:

• A viable HE community with a shared sense of purpose
• New functions at its centre – integration, caring, facilitation, curiosity, compassion and courage
• Diverse and legitimated members
• Development of demonstrative/inspirational projects
• New governance and organisation

It is also important to emphasise that the aim of the SOS is not to supersede the existing system but to work alongside it in a complementary manner. The underlying premise is that the organisation can be transformed by a process of continuous blending of the first and second operating system approaches. This is reminiscent of the indigenous concept of *Ganma* – the place where freshwater and saltwater meet and mix (Faerron Guzman and Potter, 2021). It is an attractive and compelling proposition and, in many ways, aligns with the approach Canterbury Christ Church University (CCCU) has taken for more than ten years through its *Futures Initiative*. In this next section, the strengths and limitations of the SOS approach to university repurposing will be explored by drawing on our experience.

Our second operating system: the Futures Initiative

The *Futures Initiative* was conceived as a capacity-building programme in response to CCCU's engagement with the Green Academy Change programme (AdvanceHE, 2023). Its function was understood as facilitating 'an ethical and principled response to sustainability issues' (Scoffham, 2016, p. 287). At the centre of this initiative was a small pump-priming budget that colleagues could access to support curriculum development projects (aligned with the need for demonstrative/inspirational projects). A small academic team also offered support and provided wider engagement opportunities for staff and students (residentials, seminars, campus, and co-curricular activities). The involvement and support of senior management, the governing body, and the Students Union was crucial in creating an environment in which grassroots initiatives could thrive. In addition to a diverse and legitimated membership, other key characteristics of the governance and operation of the Futures Initiative are detailed below (based on Scoffham, 2016).

* An inclusive approach that involves staff and students as partners
* A belief in voluntary and evolutionary change
* A commitment to critical and creative thinking
* Shallow hierarchies and management structures
* A culture that supports experimentation and risk taking

Since the inception of the Futures Initiative in 2011, it supported more than one hundred curriculum development projects. While some of these projects could be considered 'stand-alone and static' suggesting that their individual impact was limited, many projects acted as a catalyst for other activity and fostered wider engagement; over time a series of loose clusters and inter-connections developed. Scoffham (2016) draws on rhizomic theory to explain the way this approach led to the development of a community or network of practice (a viable community). The establishment of this large and active net-work of practice was a significant achievement and our approach has been nationally and internationally recognised. In 2017 CCCU was awarded the national Green Gown Award for 'Continuous Improvement: Institutional change' followed in 2018 by the International Green Gown Award in the same category. In 2020 we were winners of both the Advance HE Collaborative

Award for Teaching Excellence (CATE) and our institutional award for teaching excellence.

The transformational limitations of a second operating system approach in HE

Kotter's (2012) change management model suggests that once quick wins (the inspirational demonstration projects) have been established and there is a viable community of practice, a period of acceleration will follow, when changes become mainstreamed and embedded within the institutional culture. This, however, was not our experience. Universities are complex systems consisting of both a central operating system (led mainly by professional services teams) and a cluster of Schools and Faculties (led mainly by academic staff) each with its own disciplinary-infused culture and practices (Robinson and Laycock Pedersen, 2021). For a SOS to be effective it needs to engage with and bring about transformation within each of these domains. However, if like the *Futures Initiative*, it relies on a degree of voluntarism it means that assigned responsibilities in the traditional hierarchy are prioritised. It also means it is better placed to address fast variables that provide quick and visible wins leaving slow variables such as organisational culture largely unchanged. Recognising these limitations of our SOS approach, and aware of the need to increase both the scale and pace of our activity, we started to ask the question 'what must we do now?'

What we must do now: developing 'second half of life' thinking and action

It is hard to isolate the diverse influences on thinking and even more difficult to present a coherent narrative. However, I have noted previously that, compared to other organisations and institutions, universities have been relatively slow to respond to the global environmental crisis (Kemp and Scoffham, 2022). The HE sector is in the early stages of its sustainability journey and can potentially learn from looking outwards about how to adapt as it matures. The experience of 'a small country' (Wales) in making sustainable development its central organising principle of governance highlights the importance of timeliness, not just in terms of fast/slow, but also in relation to institutional maturity and what is needed at different stages. The ambitious journey to establish the Well-Being of Future Generations Act started in 1999 with the creation of the National Assembly for Wales. Initially, the focus was on securing quick small changes but as its governance structures matured, this incremental approach was no longer enough. Davidson (2020) reflects on how her level of ambition was challenged having secured school cycling lessons for children in 20 per cent of Welsh communities and being asked 'what about the other 80 per cent? This resulted in her commitment to making sustainable development 'the central organising principle of government, rather than a key theme which then had to compete with others' (p. 51). The maturation process was associated with changes in both thinking (from incrementalism to integration) and action (a re-orientation from fast quick actions to 'slow big wins').

A second important (and possibly unexpected) influence that builds on these insights about maturity is the Franciscan writer, Richard Rohr. Drawing upon the Jungian idea of the two halves of life, he argues that a different approach is needed for each with a different language. Put simply, the first half of life is about establishing an identity, relationships, and community – creating outward success. In contrast, second-half living requires an inward deepening and growth outwards beyond our comfort zone into challenge and uncertainty. It involves holding tensions, including and uniting, and understanding what matters. It is the stage of life most associated with wisdom (Rohr, 2013). The two halves are not chronological and not all people are able to embrace second-half thinking – particularly as our culture and institutions (including universities) are so firmly oriented to the first half of life. It is not about rejecting what has gone before but finding ways to 'include and integrate the wisdom of the first half of life' (p. xxvii). Thinking about universities and their engagement with sustainability, most of the advice and guidance about practice seems to be similarly oriented to the 'first half of life' – to identifying what outward success looks like, and less about 'second-half of life' thinking about how sustainability can be authentically integrated.

Returning to the Paradox Model, the task for universities could be framed as how to move from fast alignment to slow resistance as they mature in their relationship with sustainability. This is challenging territory, and I have argued previously that sustainability wisdom, a concept that integrates the cognitive, affective, reflective, and behavioural, is a vital navigational attribute (Kemp and Scoffham, 2022). Our established institutional response to sustainability, the *Futures Initiative,* explicitly valued critical and creative thinking and we have always used provocations as a way of stimulating this attribute. In this next section, I show how I have engaged with provocative myths and metaphors to develop my own understanding of 'what we must do now' and to contribute to the development of our institutional approach through the *Academy for Sustainable Futures.* Rohr (2013) recognises the unique ability of myths to 'hold together paradoxes that the rational mind cannot process by itself' positioning them as 'transrational' and pulling us into 'deep time' (p. 6). This potential to offer new insights into established paradoxes suggested myths and metaphors could offer a helpful navigational tool.

Myths and metaphors as navigational tools

Booker (2005) argues that storytelling is a deeply ingrained part of human nature and that there are certain patterns (or archetypes) that are common across time and place – these are commonly expressed as mythical tales. A myth, then can be understood as a traditional story, that often involves supernatural beings or events to explain a natural or social phenomenon. There is, he argues, a universal message. 'The first concern of stories is to show us the nature of the power of egocentricity and what it does to human beings' (p. 556). This is demonstrated most explicitly in the tragedy plot, although all plots have a 'dark' version where a main character chooses to think and act egocentrically, leading to death and destruction. One example of the tragic hero comes from the Greek myth of Icarus who fails to heed the warning to avoid flying too high or low when escaping from Crete using wings his inventor father has made for him. Icarus ends up flying so close to the sun that the wax holding his wings together melts and he falls to his death. Driven by his ego, Icarus fails to maintain a state of balance and destroys himself. This is clearly an apposite message in the contemporary context of global crisis where the warnings are clear, but humanity is failing to respond.

As well as revealing the problem (the ego), stories can also point to how things should be – they can reveal patterns of wholeness and harmony. Booker (2005) shows how myths can help us to 'see whole' (p. 253) either when a hero or heroine is 'liberated from the distortions of ego-consciousness, onto a different level which gives them a clearer understanding' (p. 254) or when they move deeper into tragedy through the 'light' characters they encounter. This paradoxical framing of the problem and offering of a resolution makes myths such a compelling form for exploring the challenges of integrating sustainability into HE contexts. It is one that I have drawn on previously to engage staff and students with sustainability in a creative manner, foregrounding emotional connection and personal responses (Wall et al, 2020). The Sustainability Stories project generated a book of fictional accounts of colleagues working in sustainability and emphasised the power of mythical tales (Field and Bonfield, 2019). Many commonly used metaphors derived from mythical tales. Based on the Greek term *metapherein* (meaning to transfer), a metaphor is an expression that describes an object or action by referring to something that, whilst very different, has some shared characteristics.

In relation to developing our institutional approach, my perception of the challenge was that it was one of classical dimensions – several different Greek classical myths came to mind, including the twelve tasks of Hercules! However, it was the mythical tale of the Trojan War, and the associated metaphor of the Trojan Horse that kept coming to mind and it was this that I have used to stimulate my thinking. Documented in Virgil's Aeneid and also referred to in Homer's Odyssey, the Trojan Horse refers to the huge wooden horse constructed by the Greeks to gain entrance into Troy. The essence of the story is that, having tried (unsuccessfully) to conquer Troy for more than ten years, the Greeks constructed the horse to hide a small group of soldiers and left it at the gates of the city. The Greek forces then sailed away. Believing the horse to be a peace offering and a symbol of their victory, the Trojans pulled it into the centre of the city. At night, the soldiers crept out and were able to reopen the city gates for the main Greek army to pour into the city and finally conquer it. The meaning of this mythical tale is debated, although it is understood to convey both cultural and political messages about the precariousness of life and the destiny of the Roman Empire. Whilst 'Trojan Horse' is a provocative metaphor as it is often used to refer to deception or trickery and applied in computing to malware that can infect computers with harmful viruses, I found it illuminating when using it to reflect on the concept of the SOS within HE contexts. In the next section, I outline some of the points of reflection.

Engaging with a resilient system: Troy was a defended city with a fortified wall and gates that meant that it was hard to infiltrate from the outside. Reading Robinson and Laycock Pedersen's (2021) paper, I came to see that their non-normative resilience lens positions the university as a structure similarly impenetrable from the outside. As they argue 'our current university systems … are resilient to change due to slow variables such as organisational and sector-wide prevailing paradigms and culture (p. 1). When faced with external drivers for change (such as the global crisis) their tendency is to add rather than integrate (fitting in the 'fast alignment' quadrant of the Paradox Model). This is the reformist position outlined in the introduction. In our case, the development and delivery of the *Futures Initiative* as an additional and complementary function of the university did not challenge the dominant operating system. The Greek army had to change its tactics when faced with a strongly defended and resilient city, and this suggests that we may need to develop a different approach as a university moving into a mature stage of engagement with sustainability.

Getting into the centre of the system. The Trojan Horse was left outside the city walls, then taken into the centre of the city. This was crucial for the Greek soldiers as it meant they could survey the surrounding territory and familiarise themselves with the context – only coming out of the safety of the horse when all the Trojans were asleep. In the original Paradox Model, I argued that the central intersection between the axes holds special significance – it is the centre of the navigational compass and a point from which to survey possibilities before taking action. Engaging with the myth reemphasises the importance of taking this position – a place Biesta (2022, p. 49) has recently termed 'the difficult middle ground'. It is difficult because it is a place of visibility and vulnerability, but it is also difficult to access. It is a carefully guarded space and, as the mythical tale reminds us, stealth may be needed to access it. Whilst deception is clearly unethical, perhaps the message we can take from the story is the importance of understanding where opportunities may exist. Here, the Greeks were able to leverage the Trojan's belief that they had been victorious and their enemy had created the horse as a tribute. Applied to the university, it may be that sector and institutional drivers can be leveraged to facilitate meaningful engagement with sustainability.

Opening the gates. Around a dozen Greek soldiers plus Odysseus were secreted in the Trojan Horse, playing a risky but essential role. They were able to open the gates and let in the rest of the Greek army, which had sailed back to Troy.

This suggests that the key role of those working in the SOS is to enlist the support of a wider army to join them in their mission (this would include those in strategic teaching, research, and operational positions).

Infiltrate the city. Once the whole Greek army was in the city, they were able to infiltrate every part and attack on multiple fronts. Few survived the attack and the Trojan culture and its people were essentially destroyed. The contemporary application of the Trojan Horse metaphor to describe harmful computer viruses emphasises this sense of complete penetration. I want to return here to my earlier point about the complexity of the university, and the limitations of the SOS approach in bringing about change of sufficient scale and pace. Like cities (in this case Troy), the university can be considered to operate as a series of interconnected subsystems. In his book, *The Ecological University*, Barnett (2018) identifies 7 'fragile' ecosystems within the contemporary university: knowledge, social institutions, persons, the economy, learning, culture, and the natural environment. Furthermore, Bessant, Robinson and Ormerod (2015) recognise four domains of activity for the university – campus, research, education, and outreach. Seen in this way, any SOS needs to find ways of penetrating each of these ecosystems and domains of activity.

Destabilising the prevailing system. The aim of the Greek army was to bring an end to the long and devastating war and achieve a decisive victory. This meant destroying the city of Troy and its people. Again, a literal application is not appropriate, but it does emphasise the need to destabilise the prevailing system and replace it with a different one. As Robinson and Laycock Pedersen (2021, p. 1) note, this is important in the context of universities. 'To repurpose a university requires us to destabilise our prevailing system, crossing a threshold into a new stable system of a "sustainable university" across all its domains'. This suggests that we need to move beyond the SOS being a permanent and complementary addition to the first system, and instead see it as the means of destabilising and ultimately, replacing it.

Our Trojan Horse? The Academy for Sustainable Futures

This playful and provocative engagement with the Trojan Horse myth helped us to develop a response to the question of 'what we must do now'. The Academy for Sustainable Futures was launched in 2022 and builds upon our existing approach (the *Futures Initiative*) whilst acknowledging the need to upscale (to move beyond an 'initiative') and to extend the academic focus (particularly in relation to research and enterprise activity). It also reflects our strengthening institutional position, in which Shaping Sustainable Futures is at the centre of our new strategic vision following widespread consultation. This, of course, has only been possible because of the foundations laid over the previous ten years through the *Futures Initiative.* The Academy sits centrally within the university

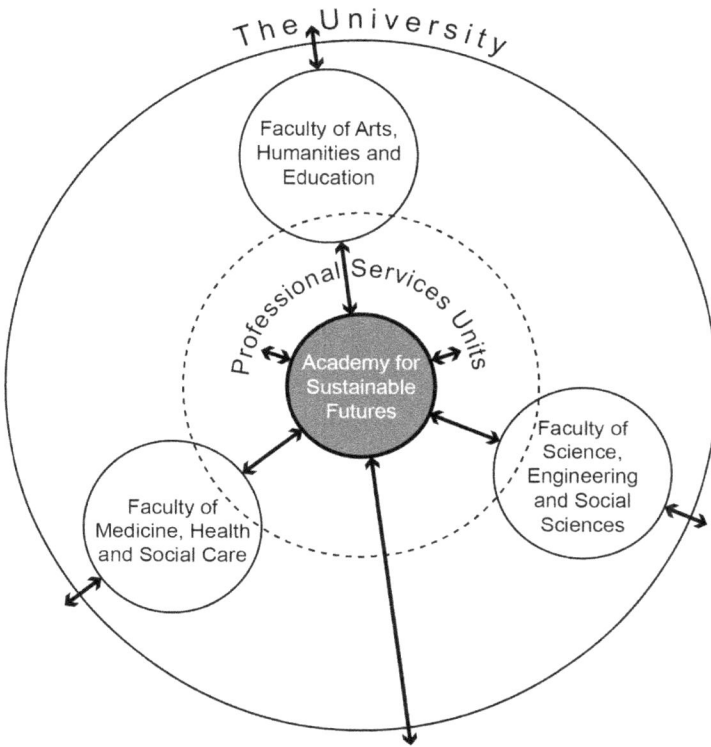

Figure 13.1 The Hub and Spokes approach of the Academy for Sustainable Futures.

and works across Faculties, Schools, and Professional Services departments to facilitate the delivery of our institutional commitment to 'shaping sustainable futures'. It operates a 'Hub and Spokes' approach, with three key elements, as illustrated in Figure 13.1.

1. Working with Schools and Faculties to build collaborative research and curriculum development networks.
2. Working with key professional services departments to develop a supportive and sustainable infrastructure.
3. Building our institutional capacity and reputation for research, consultancy, and advocacy for Sustainable Futures.

Principles of the Academy

• Promotes an inclusive and holistic understanding of Sustainable Futures.
• Acknowledges the complexities and tensions that the full scope of sustainability creates.

- Establishes understanding that sustainable processes are essential to ensuring sustainable outcomes.
- Builds academic capacity in both teaching and research, to drive curriculum change and facilitate transdisciplinary research.
- Establishes and nurtures collaborative partnerships (individuals, teams, and external organisations).
- Advocates and influences at multiple levels, internally and externally.

It is early days, and operationalising these principles is challenging and involves a constant process of negotiation. We are currently involved in exploring our shared institutional understanding of the term 'Sustainable Futures' and identifying or creating a framework that can scaffold our collective action. Like many universities, we have engaged with multiple frameworks and concepts supporting an open and inclusive approach to engagement with sustainability. Navigating and holding the complexities and tensions that arise as our second half of life unfolds is what Rohr calls 'the very shape of wisdom' (2013, p. 138)

Conclusions: extending the Second Operating System model

In universities, as in many organisations and institutions, there is a tendency to focus on what Rohr (2013) terms 'first-half of life' thinking and action prioritising quick and visible wins. The question of *how* to go beyond this and develop the type of transformational responses to the global polycrisis that Stewart, Hurth and Sterling (2022) argue are needed, has received relatively little attention. Taking inspiration from their call for both *retrospective* and *prospective* reflection, this chapter has explored the experience of one university in its journey to repurpose. It finds that whilst the concept of a SOS has many strengths, a different strategy may be needed to release its transformational intent. Once a viable and legitimate network of practice has been established within a university, there is a need to adopt 'second-half living' thinking and action and move from incrementalism (a permanent complementary SOS) to integration (SOS replacing the main operating system). Engaging with myths and metaphors can help stimulate reflection and can offer a creative navigational tool. Using the Trojan Horse metaphor it is clear that whilst the SOS still has a fundamental role to play in a mature university, there is a need to rethink its position, purpose, and practices as outlined below.

Positioning. Like the Trojan Horse, the SOS should be situated centrally rather than peripherally in relation to the existing structure.

Purpose. Like the Greek Army, the sense of shared purpose needs to extend beyond the SOS and its network of practice (the soldiers within the Trojan Horse) and be adopted at an institutional level (the wider army).

Practice. Like Odysseus, a strategic and ambitious approach is needed in practice. In addition to fostering and developing the network of practice, the

SOS needs to target and leverage key institutional drivers to integrate sustainability into governance and operational structures.

This extended model of the SOS is offered by way of a tentative response to the question of how universities can repurpose for Sustainable Human Progress.

References

AdvanceHE (2023). Canterbury Christ Church University Green Academy case study, available at www.advance-he.ac.uk/knowledge-hub/canterbury-christ-church-university-green-academy-case-study accessed 24 August 2023.

Bessant, S., Robinson, Z. and Ormerod, R. (2015). Neoliberalism, new public management and the sustainable development agenda of higher education: history, contradictions and synergies. *Environmental Education Research.* 21, 417–432. doi: 10.1080/13504622.2014.99.3933

Biesta, G. (2022). *World-Centred Education: A View for the Present.* London: Routledge.

Booker, C. (2005). *The Seven Basic Plots: Why We Tell Stories.* London: Bloomsbury.

Davidson, J. (2020). *#futuregen: lessons from a small country.* London: Chelsea Green Publishing.

Faerron Guzman, C. and Potter, T. (eds) (2021). The Planetary Health Education Framework. Planetary Health Alliance. Available at: https://drive.google.com/file/d/1wg2zJnKj-wlGN5qK8EXC0Y2lqUwMUQNV/view

Fazey, I., Hughes, C., Schäpke, N.A., Leicester, G., Eyre, L., Goldstein, B.E., Hodgson, A., Mason-Jones, A.J., Moser, S.C., Sharpe, B. and Reed, M.S. (2021). Renewing Universities in Our Climate Emergency: Stewarding System Change and Transformation. *Front. Sustain.* 2:677904. doi: 10.3389/frsus.2021.677904

Field, V. and Bonfield, C. (2019). *Not Utopia … but Maybe.* Canterbury Christ Church University.

Fraser, N. (2021). 'Climates of Capital: for a trans-environmental eco-socialism', *New Left Review*, 127 Jan/Feb 2021, pp. 94–127.

Kemp, N. and Scoffham, S. (2022). The Paradox Model: towards a conceptual framework for engaging with sustainability in higher education, *International Journal of Sustainability in Higher Education*, 23 (1) pp. 4–16. doi.org/10.1108/IJSHE-08-2020-0311

Kotter, J.P. (2012). Accelerate! *Harvard Business Review*, November, 44–58.

Robinson, Z. and Laycock Pedersen, R. (2021). How to Repurpose the University: A resilience lens on sustainability governance. *Front. Sustain.* 2:674210. doi: 10.3389/frsus.2021.674210

Rohr, R. (2013). *Falling Upward: A Spirituality for The Two Halves of Life.* London: SPCK.

Scoffham, S. (2016). Grass roots and green shoots: building ESD capacity at a UK university. In J.P. Davim and W. Leal Filho (eds.), *Challenges in Higher Education for Sustainability*. Management and Industrial Engineering, doi: 10.1007/978-3-319-23705-3_14

Sterling S. (2021). Concern, conception, and consequence: Re-thinking the Paradigm of Higher Education in dangerous times. *Front. Sustain.* 2:743806. doi: 10.3389/frsus.2021.743806

Stewart, I.S., Hurth, V. and Sterling, S. (2022). Editorial: Re-Purposing Universities for Sustainable Human Progress. *Front. Sustain.* 3:859393. doi: 10.3389/frsus.2022

Sustainable Development Solutions Network (2021). Accelerating Education for the SDGs in Universities. Available at: https://irp-cdn.multiscreensite.com/be6d1d56/files/uploaded/accelerating-education-for-the-sdgs-in-unis-web_zZuYLaoZRHK1L77zAd4n.pdf

Thunberg, G. (2022). *The Climate Book.* London: Penguin.

Well-Being of Future Generations Act, available at www.futuregenerations.wales/about-us/future-generations-act/ accessed 24 August 2023.

Wall, T., Puntha, H., Molthan-Hill, P., Kemp, N., Puntha, S and Baden, D. (2020). Stepping into Sustainable Futures: actions, developments and networks. Chapter 30 in *Storytelling for Sustainability in Higher Education: An Educator's Handbook.* Edited by Petra Molthan-Hill, Heather Luna, Tony Wall, Helen Puntha and Denise Baden. London: Routledge.

Kissing the earth

Victoria Field

When we walk like we are rushing, we print anxiety and sorrow on the earth. We have to walk in a way that we only print peace and serenity on the earth.... Be aware of the contact between your feet and the earth. Walk as if you are kissing the earth with your feet.[1]

(Thich Nhat Hanh (1991))

So many kisses on a pilgrimage
our feet kissing the earth time and again

the bread broken and kissed
softened flesh become flesh

wine glass raised to the lips
red meniscus, tannin, grape become blood

fork and spoon lifting and lowering
for kiss after kiss

our backs kissing a bed, or warm grass on a bank
fingers kissing laces on a boot, hand kissing the carved head
of a pilgrim staff relayed from here to Rome

the monk taught we should wash a teapot
with the same love and care as if we were bathing the baby Jesus or Buddha

dear teapot, just waiting to serve
and be loved in return

sweet earth kissing our feet
never counting the cost

[1] The phrase 'kissing the earth' is taken from *Peace Is Every Step* by Thich Nhat Hanh (1991, p. 28), the Buddhist monk and peace activist (1926–2022).

Reference

Hanh, T.H. (1991). *Peace Is Every Step*, Rider: London.

DOI: 10.4324/9781003286516-20

14 Good (higher) education in a fragile world

Nicola Kemp and Alan Bainbridge

There is always the possibility that edited texts can try to capture and hold too many disparate ideas, ending up with stand-alone chapters that, although valuable in themselves, offer little towards providing a cohesive whole. As editors of this complex and wide-ranging interdisciplinary volume with a focus on often intangible wicked problems, we were acutely aware of such potential pitfalls. The Introduction sought to clarify the origins and context in which this book emerged and the intellectual journey all participants were about to embark upon. As editors our task is now to take care of bringing this project to a meaningful conclusion while also highlighting future possibilities.

To return to the beginning, our starting point for the book was to ask the question of what good (higher) education – engaging in teaching, research, and scholarly activity – could or should look like in the context of contemporary sustainability concerns such as the global crises of climate change, economic and social injustice, and biodiversity loss (planetary fragility). We note Biesta (2022) argues that 'this existential question – the question how we, as human beings, exist *in* and *with* the world, natural and social ... is the central, fundamental, and if one wishes, ultimate educational concern' (p. 3 emphasis added). And yet, despite the centrality of the 'ultimate education concern' in relation to the sustainability of the planet and all its inhabitants, it is a concern that remains marginal, at best, within mainstream educational discourse.

The provocation around which we initially framed our question and brought our contributors together on what a 'good education' might be, was Kemp & Scoffham's (2022) Paradox Model. This concept proposed that there are educational paradoxes or tensions that educators (and educational institutions) need to navigate in response to the backdrop of global crisis. The basic premise is not new; Kant's educational paradox centres on the tension between freedom and restraint, Bruner (1996, pp. 66–85) articulates three 'unresolvable tensions' (individual development vs cultural reproduction; talents versus tools; particular versus universal), while Fromm (1994, p. 100) notes the contradiction of human existence 'that requires a search for solutions, which in their turn create new contradictions and now the need for answers'. Similarly,

DOI: 10.4324/9781003286516-21

Schumacher (1973) recognises that questions of meaning and purpose (particularly associated with the Humanities) generate tensions and require higher-order thinking to navigate. The contribution of the Paradox Model is to apply this educational thinking to the area of sustainability education, or education for sustainable development (ESD), an emerging field that Sterling (2014) argues has tended to run on parallel tracks to education as a discipline.

In this chapter, we reflect on the insights and contributions of the thirteen previous chapters in relation to the Paradox Model and propose a revised and extended version. We also return to answer our initial question of what good education could look like in higher education contexts.

On the importance of paradoxical thinking

Our first 'conclusion' relates to the value of acknowledging educational tensions and engaging in paradoxical thinking. In the preceding chapters, the authors demonstrate the diverse ways in which the educational tensions identified in the Paradox Model impact their efforts to engage with sustainability and how they navigate these in their practice. They also reframe these and reveal other important tensions that need to be acknowledged.

The influence of the neoliberal educational values (and resistance to/alignment with these) are explored throughout this book and Ali (Chapter 2, p. 18) powerfully frames the subsequent discussion by arguing that one of the biggest failures of Education for Sustainable Development (ESD) has been its inability to think beyond the neoliberal narrative and to 'develop, articulate and gain support for alternative and competing ideas'. Throughout the book, multiple examples of 'thinking beyond' are offered in relation to educational purposes and pedagogies. Starting from what might seem a position of 'alignment' with an instrumentalist employability agenda, Bartram & Brewer (Chapter 4) demonstrate the importance of creating reflective educational spaces to challenge from within. Their innovative applied humanities module enabled students to build their own understandings of the value (purpose) of their course by integrating narratives of utility and passion. It is perhaps in Part 3, in relation to the university as an institution that Ali's provocative challenge is most explicitly addressed. In Chapter 11 Bainbridge reframes the tension by drawing upon ecological language (tame/wild) and argues for a wilding of higher education. Wilding in this context, involves allowing spaces around the edges for creative and unpredictable educational encounters (arguably exemplified in Bartram & Brewer's approach). Pagden (Chapter 12) looks at the transition movement and considers its application to educational contexts. He revisits Ivan Illich's call to abandon formal schooling and restructure society and positions the role of Initial Teacher Education in higher education as a place where alternative ways of educating can be legitimately thought about and practiced.

In the original Paradox Model, the temporal (fast-slow) tension was understood in terms of how the demand for immediate action could be meaningfully satisfied within educational contexts. Time emerges as a central concern, particularly in terms of acknowledging the importance of slowness and the attentiveness this can facilitate (see particularly Watts, Chapter 6; Overall, Chapter 7). The value of 'slow' thinking (Kahneman, 2011), pedagogy (Clark, 2022), and purpose (Davidson, 2020) have been written about elsewhere, but this book makes an important contribution to its place in higher education. The authors in this book also remind us that authentic change takes time, whether it involves building individual staff capacity (Scoffham, Chapter 7) or transforming the organisational approach to sustainability (Kemp, Chapter 13). Other chapters highlight the significance of engaging with longer timeframes. In Chapter 9, Heath & Vujakovic, remind us that we are involved with all forms of life, past, present, and future, and that this understanding can challenge an anthropocentric worldview. Trees, for example, are demonstrated to offer interesting lessons for students in 'long-life learning' – thinking in 'tree-time' (Vujakovic, 2013). The need to integrate past, present and future perspectives is a strong message across the contributing chapters. An intergenerational perspective is of course inherent within definitions of sustainable development, where the tension is understood as one between present and future generations (WCED, 1987). Futures Thinking is acknowledged as a key dimension of ESD (QAA/HEA 2014) as it recognises the long-term impact of the decisions taken in the present. This is expressed as 'Anticipatory Competency' by UNESCO and in the updated sector guidance (Advance HE, 2021). However, the importance of connecting with the past is not typically emphasised. Our book makes a significant contribution by articulating the value of engaging with history and heritage and with other (more than human) ways of understanding time.

A spatial paradox: integrating the fragment with the whole

Across the chapters, a dominant underlying theme can be discerned related to the need to manage the tension between the fragment and the whole that we term here, the spatial paradox (foregrounding our perception of relationships in space). Different language is used by authors – part/whole, particular/ universal/, local/global – but all are effectively exploring the same challenge from different perspectives. The potential of 'fragment' to act as both a noun and a verb is also relevant as it links with a central proposition of the book (articulated most explicitly in Part 2) – that sustainability education involves pedagogies of (re)connection. The question inherent within this tension, may be expressed as 'how can a university situated in a particular place, authentically engage students with the global polycrisis?

In Chapter 6, Watts argues for a holistic pedagogy as a means of integrating whole/part relations and offers a conceptual framework for thinking about this paradox in research and teaching. The authors within this book share an understanding that there is educational value in approaching the universal (global) through the particular (local or hyperlocal). Overall, writing in Chapter 4 uses the concept of 'Synedoche' to illustrate the way in which a fragment can illuminate the whole through the practice of attentive walking. The significance of immersive (embodied) experiences of place can be overlooked in the context of the global crisis but, she suggests may offer a pathway of connection. Writing from the perspective of a teacher educator, Riordan (Chapter 10) demonstrates the value of close analysis to the details of what happens in classroom contexts when sustainability is being taught in schools. For Scoffham (Chapter 7), small-scale local interventions offer a pragmatic and authentic response for universities rather than attempting to 'fix' the world.

An affective paradox: balancing emotions of love and fear

Interestingly, the role of emotions – the affective dimension – is also a dominant theme throughout the book and is offered here as an additional paradox that educators in higher education must navigate. This tension can be understood as how to engage students with the contemporary crisis of 'unsustainability' in all its manifestations whilst promoting an ethic of love and care. In Chapter 3 Wilson dives deep into the language of love and asks the reader to imagine an education that is not just about what we know, but what we love. After all, he asks how can we care for what we do not love? And yet, he notes 'Love talk seems to have little place in any of these serious ventures: love needs the particular, the unique, the specific if it is to exist, as indeed does the world' (this volume p. 28). In contrast, calls to action to "save the planet" tend to be based on the fear of what the human consequences of planetary damage might be. The contemporary sustainability discourse he argues is an invitation to panic and be afraid. This is exemplified in Thunberg's rallying call (World Economic Forum, 2019): 'I want you to act as you would in a crisis. I want you to act as if the house is on fire. Because it is'. But, Wilson argues, being in a state of fear can prevent us from making good decisions. The response to fear is too often 'heroic and speedy', focussing on techno/managerial actions that assume an ability to control ecological interconnectivity. Encountering evidence about the nature and extent of the crisis is emotionally challenging but, clearly, universities have a role in both creating and disseminating knowledge about the state of the world. If our concerns are for a sustainable future, for the planet and individuals, then a good higher education must involve learners and teachers who are open to encountering nature, the other, and troublesome knowledge whilst being guided by love. This is also at the centre of Khovacs's argument (Chapter 5) as he explains,

> The reason is simple: as humans, we act in defence of what we love, whether land, self, community, or kin. Scaffolding a pedagogy of the environment on notions of love in response to a woundable planet is no exception: we teach from a position of self-involvement in what we love.
>
> (p. 57)

Love (and passion) for a subject is similarly central to Bartram & Brewer's understanding of educational purpose. The alternatives, Wilson provocatively argues, have the 'stench of death about them, ours and the world's' (This volume, p. 32).

A systemic paradox: revaluing fragility and resilience

In the Introduction, we situate our use of the qualifier 'fragile' and the reasons for using it in the title of this book. As we now reflect on and articulate our conclusions, we argue that fragility can itself be understood as part of a fundamental paradox, in relation to its antonym, *resilience*. To build our argument, we go back to our original term, 'fragile world'. The case has been well made in the previous chapters that human activity is not only threatening environmental systems but has introduced a new sense of fragility into humanity itself. Fragility can therefore be located in human activity; the corollary being that a change in human activity will lead to a more sustainable planet.

The question for higher education is how it can facilitate the change needed. Simon Wilson observes how higher education has become trapped by 'managerialism' and commodification, severing the learner from hope and beauty. This image of education is 'hopeless' and potentially complicit in planetary fragility. Khovacs inverts the book title (*Fragile Education in a Good World*), arguing that what may be needed is for the human activity of education itself to be understood as fragile (in contrast to the strong and resilient managerialist approach). He explores the idea of teaching as fragile, which places educators 'in a vulnerable position: does this not speak to a kind of fragility, a self-disclosing vulnerability in teaching?' (This volume, p. 57). This positioning of education as a weak and uncertain process is also considered by Bainbridge who draws on Biesta's assertion that, 'this makes the educational way, the slow way, the difficult way, the frustrating way, and we might say, the weak way, as the outcome of this process can neither be guaranteed nor secured' (2013, p. 3). Here the fragility is in relation to what can be expected to happen as a result of the educational act. This valuing of fragility is as potentially relevant to universities as individual educators.

Turning now to resilience, whilst recognising its contentious nature, we suggest that it offers a helpful counterpoint for understanding the characteristics of educational processes and systems. In Chapter 12, Pagden extrapolates ideas from the transition towns movement into an educational context and uses resilience as a lens for considering the socio-ecological connection of the school within its community. He also notes that resilience is a key principle within Assadourian's (2017) ideas of earth education (an approach that integrates ESD with Education for Resilience). Resilience is then a characteristic of a highly interconnected social system that is well adapted to its context. However, as Robinson & Laycock Pedersen (2021) argue using a nonnormative approach to resilience, a resilient social system (such as a school or university) is resistant to change. In Chapter 13, Kemp builds on this argument suggesting that universities have become very well adapted to and embedded in the neoliberal context and that their resilience is limiting their ability to change at the scale and pace needed. Drawing on the Greek myth of the Trojan War, she illustrates the ways in which universities can embrace fragility and rebuild their organisational cultures and structures around sustainability.

The Paradox Model

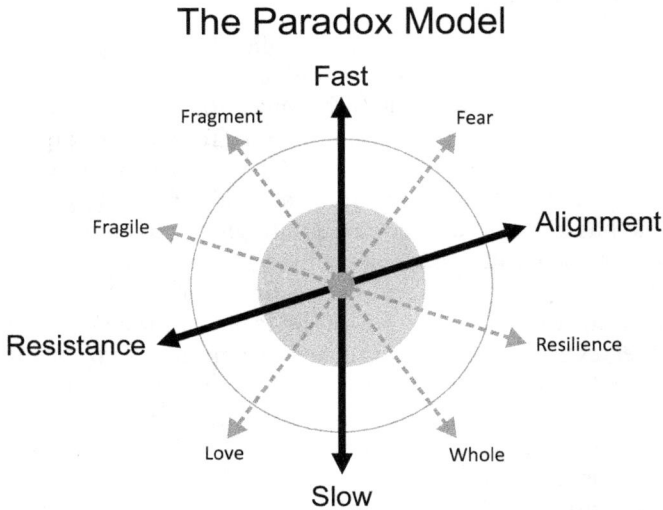

Figure 14.1 An expanded version of the Paradox Model.

This paradox causes us to revalue the concepts of fragility and resilience within sustainability education and move from a simplistic understanding that suggests that the former is the problem and the latter the solution. Instead, the terms should be seen as characteristics of educational systems that must be navigated at both an individual and organisational level. Taken together, the insights have led us to develop the Paradox Model, as illustrated in Figure 14.1.

On goodness in education

In the following section, we return to the question that has been the focus of the book – what 'good' education looks like in higher education contexts. To frame our response, we start by elaborating on the point made in the introduction about what we term 'inherent goodness'. In his book, *The Beautiful Risk of Education* Biesta (2013) draws on theology, in particular, creation stories, making a significant educational claim that 'goodness' does not reside in the act *of* creation, but rather in breathing life into what is already there. Drawing on Caputo, he notes that when 'God saw that it was good … is not a movement from non-being to being but from being to the good' (p. 67). The significance of this distinction is two-fold: first, God's creation is final and therefore the outcome is 'good'; second, it is the process of bringing to life and the uncertainty of what might happen next that is 'good'. Put simply, goodness can be a quality that characterises both the process and the product. This is important because, having reached the point of writing the conclusions, we too can say that in both senses of the word, 'it was good'.

Goodness of process

This book can be seen as an attempt to translate into academic scholarship, our thinking and understanding about what good education is. This required an attentiveness to the process that is worth detailing. Our first principle was that it should be inclusive so the call for participants was open rather than selective – this naturally created a diverse group (in terms of position, experience, and gender) who had not worked together before. Our aim was to create a diverse interdisciplinary research community. Second, we paid attention to features of time and space to meet regularly over an extended period. This involved whole writing days and monthly lunchtime seminars led by chapter authors where evolving ideas could be discussed and debated. Third, we wanted to embrace the complexity and unpredictability of what might emerge through these encounters. Finally, we wished to be sensitive to our context – this involved a deliberately parochial deep dive into one higher education setting. The process has not been straightforward, and the time taken from inception to publication has been longer than expected. To some extent the contributors have been on a pilgrimage together. By tuning in to 'our' context – our disciplines, our life histories, our professional identities – we have gained insights about the fundamental and universal (Kavanagh, 2003).

Colleagues engaged with the process noted that it was challenging and yet enjoyable to discuss emerging ideas, learn from others, and share a sense of common purpose. For most, it provided an opportunity to think about and discuss the areas of sustainability and education as a complex interrelationship, not as separate disciplines or fields of study. The process reminded us why interdisciplinary understandings are so important to develop. In the next section, we unpack and reflect on the contribution to knowledge that this interdisciplinary and collaborative process has generated. To do this, we return to the focus of the three parts of the book: rethinking educational purpose, pedagogies of (re)connection, and education as if the world mattered.

Goodness of outcome

The need for a holistic understanding of educational purpose

The titles of the chapters in Part 1 are important as they signal the diversity of ways in which the authors, coming from different disciplinary backgrounds, articulate and explore the question of educational purpose. To briefly recap, these are

For Sustainable Development
For Life
For the Future
For a Good World

What they share is a recognition that the contemporary context of environmental and social crisis requires a higher education that embraces a broader and more holistic sense of purpose than is currently the case. In Chapter 2, Ali, writing from a background in international development, offers a critique of ESD and argues that the educational challenge is to 'Swallow a World' – that is to be humble in our claims to knowledge and understanding of the world as it really is and to embrace complexity and uncertainty. Unless we at least attempt this, Ali argues that educational approaches such as ESD might only succeed in maintaining 'business as usual'. Acknowledging the limitations of ESD could enable us to recognise hidden aspects of, say, political and economic drivers. Ali provides the provocation that even revered constructions, such as the SDGs are rooted in international development architecture; mired with links to Global North versus Global South assumptions, where 'To buy into the SDG narrative is to buy into the narrative of the old colonial "civilising mission" or the white man's burden' (This volume, p. 16).

These narratives are deeply embedded and highly resistant to change, and Ali argues that education and activism should be closely aligned. Ivan Khovacs (Chapter 5) and Simon Wilson (Chapter 3) draw on theology and discuss how through this disciplinary lens, educational purpose can be conceived as a continual interaction of 'bringing to life' between the self and other – human or more-than-human. Wilson advocates for an education that awakens a sense of who we are in relation to the world and in doing so restores to ourselves and the world a more hopeful future. In contrast, Bartram & Brewer embrace the employability agenda, recognising its potential to make 'humanities insight more robust' (This volume, p. 45). They highlight the importance of maintaining a vibrant focus on the humanities to counteract what might be perceived as anti-educational drivers of economic value. Instead, the direct links made to employability (including placement experiences) enable students to recognise their particular skill set, become more confident and articulate, be better prepared for the workplace and make good decisions about their future. Based on these contributions, good education requires us to embrace diverse understandings of purpose and to interrogate our own positions.

Pedagogies of (re)connection

The chapters in Part 2 explore the question of what pedagogies might be part of a good (sustainability) education and again the initial impression is of diversity. However, they share an understanding that the educational challenge is primarily epistemological or perspectival and derives from the dominant view of human/environment separation. Pedagogic practices are therefore needed that both foster a sense of existing connection between humans and the more than human and disrupt existing ways of knowing. Watts sets up this argument in Chapter 6 drawing on a paradigm of holism. Her work with adults in higher education draws on her experience of how very young children can invite adults to be more attentive to the natural world around them. A world of wonder and playfulness that has always been there is rekindled through the experience of young children and so too is the desire to care. For Watts, the vital educational task draws on Biesta's (2013, p. 5) assertion that what is educationally important is, 'not just about how we can get the world into our children and students; it is also – or perhaps first of all – about how we can help our children and students engage with, and thus come into, the world'.

The importance of encountering the world, both physically and psychologically, is a dominant theme within this book. Overall's pedagogy (Chapter 8) incorporates hyperlocalised experiences of place through the practice of attentive walking. This involves walking slowly in edgelands, margins, liminal spaces and, everyday places, submitting to the environment without agenda. She argues that drawing the individual into communion with a fragment of the world supports an awareness of interconnections and entanglements and an ethic of care for the whole. Scoffham (Chapter 7) provides the example of the *flaneur* walking around, looking around corners and under rocks, open to serendipity to find meaning. For each, a closer observation and awareness is linked to an appreciation and care of place and all its inhabitants. The concept of entanglements is explored more fully by Heath & Vujakovic in Chapter 9. They demonstrate how their exploration of green heritage (trees and native oysters) engages students and the wider community with 'the complex entanglements between humans and non-humans, permitting radical rethinking about theories of becoming, power, and agency' (This volume, p. 111). Recognising the complexity of sustainability pedagogy, Riordan (Chapter 10) uses video-analysis to take a close look at precise pedagogical moments where issues of sustainability are encountered. Like Pagden's ITE students researching their practice (Chapter 12), this provides an opportunity for reflexivity and the honing of good educational practice. In different ways, the authors reject linear one-dimensional assumptions in favour of complex multi-dimensional nuanced insights.

Education as if the world mattered

In the final part, the message is clear – good education demands a radical move away from tinkering with current modes and practices of higher education. Our contributors consistently note how the temptation to maintain even a modified 'business as usual' approach will not bring about the epistemological and systemic changes necessary for human and planetary health. Three different conceptual approaches are presented as alternatives.

Wilding: In Chapter 11 Bainbridge draws upon ecological metaphors to explore the idea of 'wilding'. He offers an approach to education that focuses on epistemological change where to be connected to the more-than-human other and to appreciate our interconnectedness does not require physical contact, but in a shift from 'intention to attention' and the ability to sit with uncertainty, not knowing, and to be open to happenstance.

Transition: Pagden builds on the transition town movement to make an argument for a university-based 'liberal initial teacher education'. Here, the university is positioned as a space away from the constraints of neoliberalism, which has an explicit role in transitioning towards an eco-socialist future

The Trojan Horse: A provocative exploration of the Trojan Horse metaphor supports Kemp's analysis in Chapter 13 of the challenges and opportunities for universities that want to operationalise their transformational intent towards human and planetary flourishing. She offers a conceptual model based on an extension of a second-operating system (the Trojan Horse) that can infiltrate and ultimately transform the purpose and practices of the university.

A manifesto for 'good' higher education

During 2015, we worked with international photographer Mark Edwards and hosted the Whole Earth? Exhibition (part of the Hard Rain project). The project was his response to the unsustainability he saw as he travelled through and photographed our world over the course of his career. It was inspired by the lyrics of Bob Dylan's A Hard Rains Gonna Fall' which emphasise the importance of knowing and understanding a situation before you start to tell others about it. In many ways this book has been our hard rain project - it has been an opportunity to `get to know' through a collective sharing of our individual teaching, research and scholarship experiences. As a collective, we have explored our understandings of goodness and fragility. Bringing this process to a conclusion, we wish to draw on the insights from the interdisciplinary contributions provided to offer an articulation of `our song'; our manifesto for the good.

A planetary health lens. The authors within this text have made a strong case for developing a more holistic sense of purpose. Recognising the need for this to be shared, at least within our university context, we are currently exploring the potential of Planetary Health as a lens to frame our institutional commitment to shaping sustainable futures. Planetary Health is an emerging global movement and field of study focused on the mutual interconnections between environmental and human health (Faerron Guzman & Potter, 2021). A working definition for our higher educational purpose is 'to restore the health and wellbeing of the planet and ourselves now and for generations to come'. We are currently adapting the Planetary Health Education Framework (Faerron Guzman, & Potter, 2021) to reflect our specific context based on the insights from this book. Figure 14.2 offers a reworking of the model and makes two significant modifications. (1) The domains are renamed to emphasise their key focus, and; (2) The model is presented as nested concentric circles to emphasize the way in which each of the domains builds toward an integrated and holistic understanding of good education.

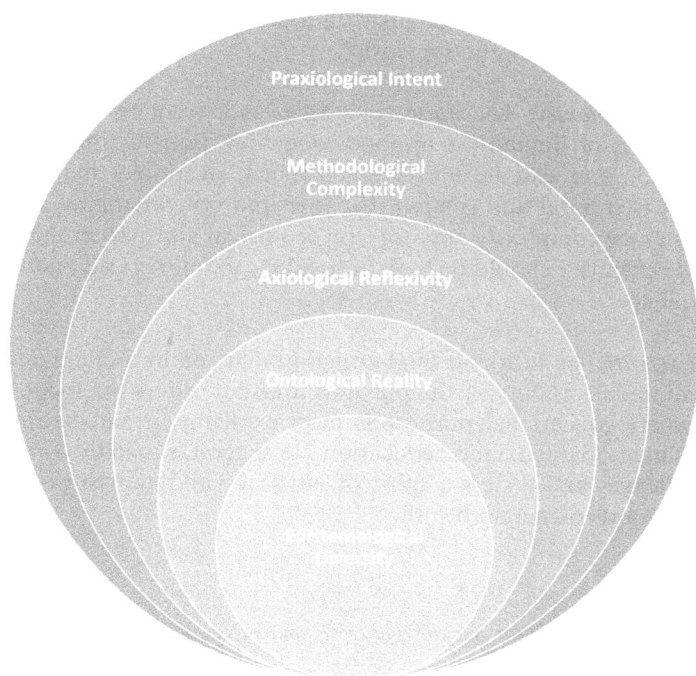

Figure 14.2 Characteristics of good higher education – a conceptual model based on planetary health.

Epistemological diversity: The authors of this book recognise that the contemporary polycrisis is rooted in a misperception that human/environment concerns are separate and disconnected. Their contributions from different religious and spiritual traditions, epistemic insights, and understandings of human/environment relationships demonstrate the potential of integrating diverse knowledge (epistemic) traditions.

Ontological reality: Transdisciplinary understanding about 'the state of the world' that recognises the boundaries of human and environmental systems, and their interconnections is inherent within our approach. The value of the arts and humanities as well as science and technology are acknowledged as fundamental to understanding fragility. The starting point for engagement could be the question 'what can your discipline contribute to our understanding of Planetary Health?'

Axiological reflexivity: The social, economic and political origins of human and planetary fragility are explored throughout the book alongside their unequal outcomes. Our understanding of good higher education foregrounds values, ethics, rights and responsibilities. Embracing historical depth and global breadth, such an approach aligns with a decolonisation of the curriculum and with the need to 'swallow a world'.

Methodological complexity: The authors of this book recognise that 'tame' or linear thinking is of limited use in good education. Equally, the idea that a solution 'works' and can be universally applied, overlooks the importance of context (temporal and spatial). Creative and critical approaches are needed that embrace complexity and draw on ecological and system thinking. Reflection and reflexivity should be built in to adapt to unintended and unexpected consequences.

Praxiological Intent: The fifth and final domain builds on and integrates the knowledge and understanding of the previous four but moves the focus onto action. It is where educational purpose is most explicitly articulated and is intrinsically change-oriented. Good education, we suggest, explicitly confronts real-world challenges and fosters a spirit of advocacy and activism to bring about change towards planetary health.

References

Advance HE (2021) *Education for Sustainable Development Guidance*. Available at: bit. ly/45CDQtf

Assadourian, E. (2017) 'EarthEd: Rethinking Education on a Changing Planet', in Worldwatch Institute, *EarthEd: Rethinking Education on a Changing Planet*, Washington, Covelo and London: Island Press.

Biesta, G. (2013) *The Beautiful Risk of Education*, London: Routledge.

Biesta, G. (2022) *World-Centred Education: A View for the Present*. London: Routledge.

Bruner, J. (1996) *The Culture of Education*. Cambridge MA: Harvard University Press.

Clark, A. (2022) *Slow Knowledge and the Unhurried Child. Time for Slow Pedagogies in Early Childhood Education*. London: Routledge.

Davidson, J. (2020) *#futuregen: lessons from a small country*. London: Chelsea Green Publishing.

Faerron Guzman, C. & Potter, T. (eds) (2021) The Planetary Health Education Framework. Planetary Health Alliance. Available at: bit.ly/3R8ej70

Fromm, E. (1994) *On Being Human* (Edited by R. Funk). London: Bloomsbury.

Kahneman, D. (2011) *Thinking, fast and slow*. Farrar, Straus and Giroux.

Kavanagh, P. (2003) *A Poet's Country: Selected Prose*, A, Quinn, (ed.) Dublin: Lilliput Press.

Kemp, N. & Scoffham, S. (2022) The Paradox Model: towards a conceptual framework for engaging with sustainability in higher education. *International Journal of Sustainability in Higher Education*, https://doi.org/10.1108/IJSHE-08-2020-0311.

QAA/HEA (2014) Education for Sustainable Development Guidance. Available at: bit. ly/3EpNkw3

Robinson, Z. & Laycock Pedersen R. (2021) How to Repurpose the University: A Resilience Lens on sustainability governance. *Front. Sustain.* 2:674210. doi: 10.3389/frsus.2021.674210.

Schumacher, E.F. (1973) *Small Is Beautiful; Economics as If People Mattered.* New York: Harper&Row.

Sterling, S. (2014) Separate tracks or real synergy? Achieving a closer relationship between education and SD post 2015. *Journal of Education for Sustainable Development* 8(2) 89–112.

Vujakovic, P. (2013) Phytobiography: an approach to 'tree-time' & 'long-life learning'. *Arboricultural Journal*, 35(2), pp. 134–146.

World Economic Forum (2019) 'Our House is on Fire' 16-year-old Greta Thunberg wants action. Available at: bit.ly/47WXRfN

World Commission on Environment and Development (1987) *Report of the World Commission on Environment and Development: Our Common Future.* Oxford: Oxford University Press.

Endword

Victoria Field

Book is a noun that is also a verb. A book is a thing, like a medieval religious relic, that does things. A book is organic in form and content. It grew into the English language, with its leaves and branches, from *boc,* the old German for beech tree. Beech wood tablets were used as writing material before paper was invented. Beech is considered a native tree in Southern England and *pannage*, the grazing of beech mast by pigs, a custom carried out for centuries for the nourishment of all. Touch your book. Sniff your book. Kiss your book. Eat your words. Some cultures consider all books sacred. Some of us attribute personhood to trees. When the Russian poet, Alexander Pushkin was dying after a fatal duel, he waved to his bookshelves and said, 'Goodbye, my friends'.

Index

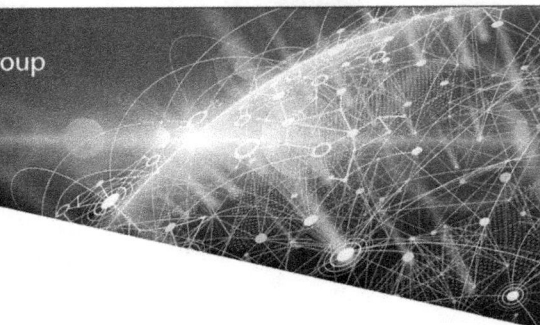

For Product Safety Concerns and Information please contact our EU
representative GPSR@taylorandfrancis.com
Taylor & Francis Verlag GmbH, Kaufingerstraße 24, 80331 München, Germany